Atmospheric Constraints in HF/UHF/Satellite Communication in the Indian Sub-Continent

Atmospheric Constraints in HF/ UHF/Satellite Communication in the Indian Sub-Continent

Dr. V. Gunasekhar Reddy
MSc., Ph.D, FIETE, MIEEE, CE, PGDBA, CNA
Deputy Inspector General of Police & Addl. Director, Police
Commissioner, A.P. Police, Hyderabad, Andhra Pradesh

SERIALS PUBLICATIONS
NEW DELHI (INDIA)

First Published – 2010

ISBN: 978-81-8387-360-4

Published by

SERIALS PUBLICATIONS
4830/24, 1st Floor, Ansari Road
Darya Ganj, New Delhi-110002 (India)
Phone : 23245225. Fax : 91-11-23272135
E-mail: serials@satyam.net.in

DEDICATED TO

Lord Sri Venkateswara & My Parents

CONTENTS

Foreword *ix, xi*

Acknowledgements *xiii*

1. Introduction 1

2 Wireless Communication Networks in Government Organisations in India 18

3 Atmospheric Radio Noise Measurements in India- at Low Latitudes 41

4 Analysis of Radio Signal Fading in HF Circuits 66

5 Mobile Communications-UHF Signal Path Loss Measurements 95

6 Tropospheric Scintillation Studies on Satellite Communications Links & Scintillations Predictions Using Radiosonde Data 116

7 Attenuation of Insat 3B Ku Band Satellite Communications Signals Due to Rain 145

8 Attenuation of Ku & Ka Band Satellite Communications Signals due to Water Vapour Absorption-Predictions using Radiosonde Data 175

9 Summary and Conclusions 194

References 203

FOREWORD

The increasing demand for satellite communications in commercial enterprise, GPS Navigation satellites, and DTH services has led to spectral congestion and now India is looking forward for the use of higher frequency bands. With the unprecedented explosion in mobile communications in India, the designers should investigate the characteristics of urban radio channels and define optional locations for the Radio ports and provide performance predictions to service providers and to the subscriber community. India will soon be adopting Digital video broadcasting through 3G & 4G techniques which requires robust decoding techniques in the urban environment. Hence knowledge of radio wave propagation in general and mobile communications in particular is very much needed.

This book is having excellent amount of information about the propagation conditions in tropics. Propagation modeling in tropics is difficult due to large variation of temperature and humidity leading to sudden changes in the atmospheric refractive index. As all our communication signals will have to traverse through troposphere, signals suffers from molecular absorption from several GHz to several Hundred GHz primarily caused by water vapor and Oxygen. Water vapor absorption which is very high in the tropics can not be easily predicted because of its high variability in space and time. Similarly, the attenuation due to rain (and Hydrometeors in general) can also be severe in the tropics. In addition to attenuation, rain can cause depolarization, scatter, amplitude and phase scintillations, Bandwidth coherence reduction etc. Thus attenuation due to water vapor and rain is a primary destabilizing parameter with Satellite radio systems in the tropics and some of these tropics are covered for professional use in this book.

This book is a good for students and researchers working in the area of radio wave propagation and to post graduate students and radio Engineering professionals. The presentation is in simple and elegant English. I wish all the best to the author.

Prof. S. Vijaya Bhaskara Rao
S.V. University, Tirupati.

FOREWORD

The Twentieth Century was dominated by an Explosion in the area of Radio Communications. Beginning with the breakthrough achieved by Marconi who demonstrated Trans-Atlantic Radio link on 12th December 1901, the century saw HF communications reach the pinnacle during the second World War when HF via Ionosphere was the only option available for long distance Wireless links and Broadcast. Very soon, VHF and Microwave links came up due to advances in generating High Power Microwaves. This era transcended to the saga of Satellite communications which eventually caters to a host of Broadband applications such as color television, High Speed Data Transfer, GPS application, Command and Surveillance etc.

However the Tropics are crowded with densely populated developing Countries where nearly 70% of the World population is living and these Countries still depend significantly on HF communications and this book gives a detailed analysis of the problems we have to contend with HF in the tropics. When we extended to frequencies in the UHF and SHF bands, especially those systems using satellites we face an entirely different set of problems. As we live in the Troposphere, all our communication signals will have to traverse through this region. Molecular absorption from several GHz to several Hundred GHz is primarily caused by water vapor and Oxygen. While Oxygen absorption can be predicted accurately, water vapor absorption which is very high in the tropics can not be easily predicted because of its high variability in space and time. Similarly, the attenuation due to rain (and Hydrometeors in general) can also be severe in the tropics. In addition to attenuation, rain can cause depolarization, scatter, amplitude and phase scintillations, Bandwidth coherence reduction etc. Thus attenuation due to water vapor and rain is a primary destabilizing parameter with Satellite radio systems in the tropics and some of these tropics are covered for professional use in this book.

With the unprecedented explosion in mobile communications in India, the designers should investigate the characteristics of urban radio channels and define optional locations for the Radio ports and provide performance predictions to service providers and to the subscriber community. A chapter in this book deals exclusively with these topics and presents results of dedicated observational schedule planned and conducted by the author.

This book is a good mix of simple and complex problems in communications that can cater to a wide spectrum of readers including Post Graduate students and radio Engineering professionals. The presentation is in simple and elegant English and I am sure will find favour with a large clientele

Dr. B.M. Reddy

ACKNOWLEDGEMENTS

I express my deep sense of gratitude and profound respects to my External Guide Dr. B.M. Reddy, Scientist, NGRI, Hyderabad for accepting me as his student and for his inspiring and excellent guidance and encouragement through out the course of my research work and without his vision and directions, I could have not turned out this task in this manner. I have acquired the status of a student of Dr. B.M. Reddy.

I express my great pleasure to record my gratitude and thanks for my Teacher and Research Supervisor Dr. S. Vijaya Bhaskara Rao, Professor, Dept. of Physics, Sri. Venkateswara University, Tirupati for his continuous encouragement, excellent guidance, benevolent and affectionate disposition in completion of the book. His keen interest, and critical comments helped me to understand the intricacies and finer points of the subject matter.

I express my gratitude and thanks to Dr. D.R. Lakshmi, Emeritus scientist, Osmania University, Hyderabad, Dr. Sarkar, Dr. Dabas and Dr. Prasad of National Physical Laboratory, New Delhi for their valuable suggestions and support.

I express my thanks to (Rtd.) Prof. J. Lakshmana Rao, Ph.D., Principal, Former & Professor of Physics, Sri. Venkateswara University, Tirupati, Prof. V.V.R. Narasimha Rao, Dept. of Physics, S.V. University, Tirupati, Prof. B.C. Venkata Reddy, Vice Principal (Rtd.), Prof B. Krishnamma, Dean, SMPS (Rtd.), S.V. University Tirupati, Prof. V. Sundararaja (Rtd.), Department of Physics, S.V. University, Tirupati and Prof. S. Buddudu, Prof. D. Narayan Rao & Dr. V.V. Subba Reddy for their encouragement.

I express my thanks to Prof. M. Krishnaiah, Prof. Ashok Kumar, Late Prof. Raja Reddy, Dept. of Physics, for their support during my Ph.D., work.

I record my sincere gratitude to Dr. M.Y.S. Prasad, Former Director, MCF, ISRO, Hassan, for his valuable suggestions, encouragement and support in data collection at MCF, Hassan for the thesis work. I also thank his team members Sri. B.V. Kannade, Group Director, Sri. S.K. Gururaj; Group Director, Sri. K. Jayaram, of MCF, Hassan for their help.

I take this opportunity to express my deep sense of respect and gratitude to Mrs. B.M. Reddy and Mrs. S.V.B. Rao for their support and encouragement during the course of this work.

It is a pleasure to record my appreciation to my wife Smt. D. Sarasavani, who was the driving force and gave me strength to complete this work and to our beloved children Chy. V. Geetha and Chy. V. Srinivas.

I am highly indebted to my parents Sri. V. Kannaiah Reddy and Smt. V. Yasodamma, my father-in-law Major N. Duraiswamy and Mother-in-law Smt. R. Chintamani, my sisters & my brother for their moral support and continuous encouragement throughout this task.

CHAPTER–1

INTRODUCTION

OUR ATMOSPHERE

A radio communication system consists of a transmitter, a transmitting antenna, the propagation medium, a receiving antenna and a receiver. Our interest here is the propagating medium, which in the terrestrial atmosphere can be either the neutral atmosphere or the ionized atmosphere. The intervening middle atmosphere between the tropo-pause and the ionosphere is of little consequence in radio propagation as the Radio Refractive Index in that region is essentially unity, like in vacuum. While the total mass of our atmosphere is about 5×10^{18} Kg, half of this mass lies below 5.5 km altitude and more than 99 percent below 30 Km. From radio communications point of view, the relevant regions are the lowest part of our atmosphere, called the troposphere extending upto 17 Km in the equatorial region, but only upto 8 Km altitude in the polar regions, and the ionosphere from 60 Km altitude to more than a thousand Km altitude. A propagating radio wave is a flow of energy through the atmosphere in the form of an electric field E and a magnetic field H with both E and H as well as the velocity of propagation being mutually perpendicular for a plane electromagnetic wave. The orientation of the electric field component of the propagating radio wave decides the polarization. The atmosphere can be classified into several regions based on the thermal structure (figure 1). Troposphere is characterized by an average lapse rate of about 6°C per Km while the stratosphere presents a permanent temperature inversion due to the significant presence of ozone. This phenomena continues upto stratopause (50 Km) from where again the temperature decreases with altitude. This decrease in the mesosphere depends on the ratio CO_2 to O_3, because CO_2 decides the cooling rate while O_3 represents heating. The lowest temperature in our atmosphere is obtained at mesopause (85 Km) and the temperature profile upwards of mesopause exhibits a steep positive gradient and this region is known as the thermosphere. Obviously, the energy input in the thermosphere is the solar Extreme Ultraviolet radiation (and some X-rays), but the steep increase in temperature above mesopause is also due to a bottleneck for energy loss due to non-equilibration of energy between the available degrees of freedom. For example, atomic species with high kinetic energy may not share that energy with molecular species which are the main cooling agents through radiation into space from infra-red active molecular vibrations.

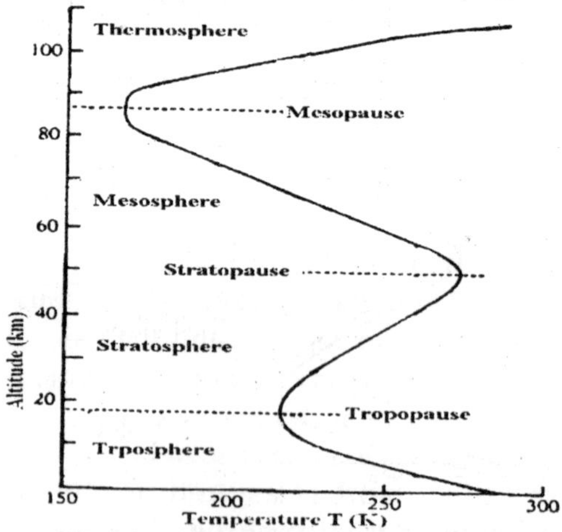

Figure 1: Temperature–altitude Profile for Earth. The Curve Shown represents the Mean Structure for Latitude 40° N During June. (Data from Houghton, J.T. *The Physics of Atmosphere*, Cambridge University Press, 1977).

The Troposphere

Troposphere by far is the most important segment of our atmosphere for every sphere of human activity including radio communications, for any mode of wireless communication has to traverse through troposphere. The troposphere is dominated by a variety of circulation systems on a wide spectrum of spatial and temporal scales. The planetary and synoptic scale motions give rise to zonally symmetric circulations such as Hadley cells, monsoons, cyclonic and anticyclonic phenomena that are so important for our very existence. The radioactive transfer and water vapour phase transformations add fuel to the atmospheric dynamics. Thermal structure of troposphere is important from radio communication point of view also. The path radio waves take, is guided solely by the Radio Refractive Index (RRI) distribution. The modification of this RRI by the water vapour in the lower troposphere plays a dominant role in microwave communications. The dry adiabatic lapse rate in the troposphere is 9.8^{0}C per Km. However, the presence of water vapour modifies this as the water vapour rises up, cools adiabatically, condenses and dumps the latent heat at different levels. This infusion of latent heat of condensation at different heights is responsible for modifying the lapse rate from the dry adiabatic lapse rate to the more usual 6°C per Km. There are a number of occasions when the standard lapse rate is modified in such a way that temperature decreases with height. These are known as temperature inversions, which dramatically increase the stability of the atmosphere, inhibiting vertical mixing and intensifying air pollution problems. Such inversions also modify RRI profiles very critically causing anomalous radio propagation problems.

The Ionosphere

Free electrons and ions constitute a very minor fraction in our atmosphere and in fact are much lesser than one percent even in the densest parts of the ionosphere. But they play a crucial role in Radio propagation and in many geophysical phenomena such as geomagnetic field variations, auroras, whistlers etc., But it was the primacy of HF communications during most of 20th century that fuelled intensive research activities on the physics and morphology of the ionosphere. Most of the sun's ionizing radiations are absorbed at levels above 60 Km. The rapid increase in mean free path with altitude in itself favors longer life time for charged particles. In addition, internal energy considerations make molecular ions more vulnerable to destruction than atomic ions (for example through dissociative recombination) and these atomic ions (mainly O^+) are in greater relative abundance at higher altitudes and hence electron number densities of the order of 2×10^6 per cubic centimeter are found in the F-region of the ionosphere around 300 to 400 Km altitude, though the production maximum is near 160 Km. A simple classification of the ionosphere is D-region from 60 to 90 Km, the E-region from 90 to 130 and the F region above 130 Km with a peak in the 300 to 400 Km range. The ionosphere above the F-peak, known as topside ionosphere merges gradually into the so called protonosphere at altitudes above 1000 Km.

RADIO PROPAGATION MECHANISMS IN THE ATMOSPHERE

The electronics and hardware components of HF/VHF/UHF propagation systems have advanced by leaps and bounds in the last few decades. So, to realize the full potential of such sophisticated systems, it is essential to understand and quantify the limitations imposed by the intervening medium, the atmosphere. Most of the problems discussed in this thesis focus on this objective so that appropriate tolerances are provided in the system.

Tropospheric Propagation

Essentially all radio systems we use will have to pass through troposphere, for that is where we live. There are several distinct types of links: Line Of Sight (LOS) systems, Over The Horizon (OTH) systems and Satellite Communications (SATCOM).LOS systems are typically operated in microwaves by using dish antenna mounted on tall towers to have an extended range. But RRI gradients, especially in low latitudes, further extend this range due to refraction as explained later. For all practical purposes, the troposphere may be considered to consist of dry air and water vapour from RRI considerations. Air molecules do not possess a permanent dipole moment, but a small dipole moment is induced by the external electromagnetic field of a passing radio wave in molecules in its sphere of influence. Water vapour molecules do possess permanent dipole moments which align with the external e-m field and add to the polarization of the medium. With the aid of constants derived from experiments, the refractivity N can be written as

$$N = \frac{77.6}{T}\left(P + \frac{4810e}{T}\right) \qquad \text{—— (1.1)}$$

Where $N = (n - 1)10^6$, n being the RRI of the medium, p is the dry atmospheric pressure and e, the water vapour pressure in millibars and T is the absolute temperature.
Equation (1.1) can also be written as:

$$N = 77.6\ p/T + 3.73 \times 10^5 \times e\ /\ T^2 \qquad \text{--(1.2)}$$

It should be obvious from the above equations that water vapour plays a dominant role in tropospheric refractivity. By differentiating equation (1.2) we get

$$dN/dh = 77.6\ [1\ dp/Tdh - (p/T^2 + 9620e/T^3)$$

$$dT/dh + 4810\ de/T^2dh]\ m^{-1} \qquad \text{--(1.3)}$$

In the atmosphere, dp/dh and de/dh are always negative. Also, usually dT/dh is also negative in the troposphere; But if it becomes positive, dN/dh in equation (1.3) will assume a large negative value and thus leading to super refraction and ducting. By substituting standard (average) gradients of pressure, temperature and water vapour at ground level for temperate latitudes in equation (1.3), we get

$$dN/dh = -4 \times 10^{-2} \text{ per meter}$$

$$\text{or} \quad -40 \text{ N per kilometer} \qquad \text{--(1.4)}$$

This value of –40 N per Km, known as initial refractivity gradient will extend the radio horizon equivalent to a straight line path on an imaginary Earth with a radius 4/3 times the true radius of the Earth. Observation in India show that for an over whelming percentage of time, the initial refractivity gradient is much more negative and super refraction is very common *(Sarkar et al., 1985)*. In fact, even values more negative than–157 N per Km when ducting takes place are not uncommon. The problems caused by high temperatures and high humidity values in radar microwave propagation are well known *(Hall, 1979)*. Hydrometeors in general and rain in particular, because of the large dielectric constant of liquid water, produce heavy displacement currents at frequencies above 10 GHz and hence large attenuation. The most important effects of hydrometeors on microwave propagation can be summarized as below *(Ippolito, 1981):*

(a) Attenuation caused by dissipation of radio wave energy as heat

(b) Scatter, resulting in loss in the desired direction and also causing interference to users in other directions.

(c) Depolarisation due to the non-spherical nature of rain drops.

(d) Rapid amplitude and phase scintillations caused by equivalent multipath propagation.

(e) Antenna gain degradation due to phase dispersion of ray paths reaching the Antenna.

(f) Bandwidth coherence reduction especially in digital systems involving carriers spanning over large channel bandwidths.

One major problem restricted to tropics is the very large diurnal variability in temperature (44°C max. and 21°C mini.) and humidity levels that cause design problems in LOS and troposcatter communication systems. Figure 2 shows how steeply refractivity changes with relative humidity at higher temperatures which are typical of low latitudes. Such dispensation necessitates use of smaller aperture antennas with higher transmitter powers.

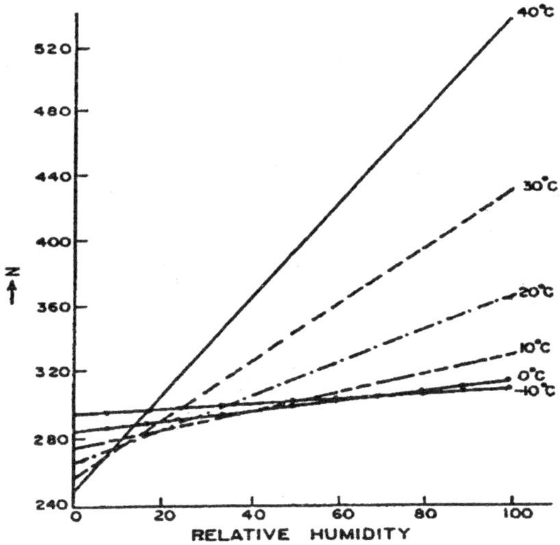

Figure 2: Calculated Values of Radio Refractivity Plotted against Relative Humidity at Different Temperatures, Demonstrates Low Dependence on Relative Humidity at Low Temperatures. N is Sensitive ar High Humidities and Temperatures (Tropics)

Radio wave attenuation due to gaseous absorption is also of special significance at low latitudes. The first three absorption bands for gases are centered at 22.2 GHz for water vapour and at 60 GHz and 118.8 GHz for molecular oxygen. Oxygen bands are not considered in this work as no services are planned in India in these frequency bands in the near future and in any case, Oxygen distribution being invariant, it has a fixed morphology. Water vapour concentrations are not only high in the tropics, but also are highly variable and ready-to-use tables are necessary in the frequency bands of concern so that appropriate margins can be given by link designers. An atlas was brought out by the National Physical Laboratory, New Delhi *(Sarkar et al., 1982)* giving water vapour profiles in the troposphere at a number of places in India. The importance of water vapour attenuation becomes obvious if it is realized that intense rain cells are only a few Km in size while radio waves in terrestrial links have to pass through hundreds of kilometers of water vapour-laden troposphere.

Depolarization due to the non-spherical nature of rain drops restricts frequency re-use systems and is compounded by multipath propagation in terrestrial links. Preliminary experiments conducted over New Delhi–Agra links showed that even for moderate rains (50 mm/hour), depolarization can bring down the reliability considerably at 19 GHz. The cross-polarization discrimination (XPD) required for high reliability systems can be as large as 20 dB. It has been found to be more practical to derive empirical relationships which are region-specific and circuit and frequency-specific including elevation angle, canting angle etc.,

While Tropo-scatter propagation is now restricted to some exclusive users, the scatter communication subject is a good example of our poor understanding of the troposphere vis-à-vis radio propagation. There is no agreement between theory and observations. Broadly, the transmission loss increases as the cube of frequency up to 3 GHz. Beyond this frequency index becomes unity.However, instances abound when the attenuation even decreases with frequency and there is no unique way that this anomaly can be explained.

IONOSPHERIC PROPAGATION–LOW LATITUDE PROBLEMS

The Ionosphere is a dispersive medium, that is, velocity of propagation depends on the frequency. When an electromagnetic wave enters the ionosphere, its propagation is governed by a complex refractive index n_i given by

$$n_i = \mu - i\chi \qquad \text{—(1.5)}$$

where μ and χ are the real and imaginary components. The refractive index is described in full by the Magneto-ionic theory developed by *Appleton and Hartree (Ratchiffe, 1959)* and generalized by *Sen and Wyller (1960)*. The refractive index is a function of the electron density N_e, the Earth's magnetic field B and the collision frequency ν of electrons with neutral and ionized constituents of the atmosphere. Thus, because of the Earth's magnetic field, the ionosphere is bi-refracting and the radio wave splits into two characteristic waves known as ordinary (O-wave) and extra-ordinary (X-wave) waves which propagate with different characteristics. The refractive index in the ionosphere is less than unity and it decreases with height upto the peak of the F-layer. Its phase velocity increases while its group velocity decreases and the wave front deviates more and more from normal as it penetrates deeper. The wave may eventually get reflected through this process of successive refraction. At vertical incidence, the condition for ordinary wave reflection (which is the important one as it suffers lesser abrorption) from a layer of electron density N (m^{-3}) for a propagating wave of frequency f (Hz) is satisfied when

$$f = \sqrt{\frac{Ne^2}{4\pi m_e \varepsilon_0}} \qquad \text{—(1.6)}$$

where e is electronic charge, m_e the mass of the electron and ε_0 is the permittivity of free space and is $1/36\pi \times 10^{-9}$ Faradays per meter.

Equation (1.6) reduces to $f = 9\sqrt{N}$ —(1.7)

As the radio wave passes through the ionosphere, it suffers attenuation and this ionospheric absorption is an important parameter to be taken into account while designing HF links or in HF broadcasting. The radio wave in its path sets free electrons into oscillation. These oscillating electrons collide with neutrals and ions in the ionosphere and part of the electron energy is dissipated. The attenuation in the D region where the wave undergoes little diffraction is known as non-deviative absorption while the attenuation in the reflecting layer is the deviative absorption. Non deviative absorption is very heavy at lower frequencies, but the deviative absorption can be prohibitively large close to the critical frequency of the layer.

The low latitude ionosphere does provide a number of advantages to aid HF communication. They are (i) large frequency range because the F-region electron densities are very high (ii) Minimum outages due to magnetic storms and other particles events as the low latitude ionosphere is secured from solar particle invasions due to the horizontal nature of the Earth's magnetic field around the equator. Thus, the disruptions due to solar disturbances essentially are limited to the Short Wave Fadeout (SWF) events which are caused by intense X-ray emissions during solar flares and last much less than an hour. The advantage offered by high electron densities is obvious if we consider the convenience of using the higher end of the HF spectrum by way of lesser non-deviative absorption and lower Atmospheric Radio Noise (ARN) as explained in Chapter-3. Most of the countries located in low latitudes which also happen to be the so called developing countries have been exploiting these advantages and stressing the HF spectrum to its limits as this mode of communication is the least expensive.

However, the low latitude ionosphere displays several characteristics that affect the reliability of HF communications and some of them have been discussed in detail by *Lakshmi et al., (1979)*. The problems can be listed as below:

(a) Atmospheric Radio Noise (ARN) is larger at equatorial latitudes, as the equatorial belt experiences overwhelmingly intense thunderstorm activity. The problem is especially serious during night time as the ARN is known to be higher. Also the Optimum Working Frequencies (OWF) during night time are very Low where the ARN is prohibitively large.

(b) A bane of low latitude ionosphere is the large diurnal range of electron densities dependent on local time. The physics behind this vulnerability is easy to understand. The total plasma content in the magnetospheric field tubes is very low at low latitudes compared to mid and high latitudes. This causes a rapid emptying of plasma near equatorial field tubes soon after sunset and most of the electrons even in the F-region are destroyed. Consequently night time frequency allocation in HF band becomes critical and to avoid too many frequencies (and hence antenna systems), a practical way is to settle down for a very low frequency near the lower boundary of HF spectrum where one has to contend with higher ARN levels.

(c) Day-to-Day variability in ionospheric parameters, especially in $f_0 F_2$ values is a major problem in low latitudes and presents a challenge to ionospheric forecasters. This random variability apparently unrelated to any specific solar/geophysical event, also known as geophysical/ionospheric noise defies modeling efforts and can at best be taken care of crudely by developing statistical pictures *(Rush et al., 1974 and Aggarwal et al., 1979)*.As the Total Electron Content (TEC) of the ionosphere is related to the f_0 F_2 trend, similar variations plague the stand alone GPS performance as it is difficult to predict the TEC variability. India is launching a Space Based Augmentation system known as GPS and Geo Augmented Navigation (GAGAN) for aiding Civil Aviation. One of the major concerns remains to be the ionospheric error which can not be easily modeled at low latitudes *(Sarma et al., 2000)*. A large number of studies were conducted in India which showed no correlation between TEC and solar or magnetic activities *(Rama Rao et al., 1981 and Dabas et al., 1984)* and this area needs further investigation.

(d) Problems due to ionospheric irregularities/ scintillations are of serious nature in HF Communications, broadcasting and Trans-ionospheric radio systems, including navigation and positioning systems *(Bhattacharya et al., 2000)* and other satcom applications even upto 15 GHz. The broad consensus reached through a number of excellent efforts on possible physics in generation of these irregularities is as follows: Post sunset plasma bubbles in the equatorial region are generated in association with the non-linear evolution of plasma instabilities (Rayleigh–Taylor type). Electron density depletions or plasma bubbles generated from such instabilities rise from the bottom side to the top side of F-region.

Steep gradients at the edges of these bubbles help to generate the smaller scale-size irregularities which produce intense scintillations even above 10 GHz. Intense scintillations can cause unacceptable bit error rates in digital systems and seriously affect trans-ionospheric systems in the equatorial region, especially in equinox months during high solar activity years. Several attempts were made from India to develop the morphology of low latitude scintillations *(Krishna Murthy et al., 1978; Rama Rao et al., 1980; Dabas and Reddy, 1989)*. During the early seventies of last century, when scintillation measurements were made only in VHF and UHF, a distinctly decreasing trend was noticed with increasing frequency. From observations at ten frequencies between 138 MHz and 2.9 GHz transmitted from one satellite, *Freemouw et al., (1978)* showed a consistent wave length dependence of $\lambda^{1.5}$ for scintillation index values not exceeding 0.6. By extending such wave length dependence to microwave frequencies up to 12 GHz, it was concluded that scintillations will not pose a problem above a few GHz. However, this state of complacence was short lived when several workers reported rather intense scintillations even around 10 GHz during high solar activity periods.

While the spread F in the equatorial region which has been historically linked to fading in HF systems, has similar origins as to irregularities described above, there seems to be a very distinct difference. The evidence from correlations of scintillation occurrence and spread F show that it is only the equatorial range spread F that is associated with scintillation activity and not the frequency spread F *(Rastogi, 1980)*. Thus, the use of available spread F maps to study trans-ionospheric scintillations will lead to misleading results. They can be used only to interpret fading in Ionosphere-reflected communications.

(e) Large horizontal gradients in the F-region electron densities associated with the Equatorial Ionisation Anomaly (EIA) can impose limitations in giving frequency predictions to North-South HF links in low latitudes. Figure 3 shows clearly the horizontal gradients associated with EIA at F-peak and above from Aloulette-1 topside sounder data. Detailed description of problems posed by such spatial gradients is given by *Lakshmi et al., (1979)*. The Maximum Useable Frequencies (MUF) may vary by as much as 10 MHz. (Figure 4) for electron density horizontal gradients of 2 electrons/CC per metre. In addition, the response of the EIA to geomagnetic disturbances is not unique through inhibition seems to be more common. *Abdu et al., (1993)* studied the global response of EIA to magnetospheric disturbances using ionosonde data from different longitude sectors during the SUNDIAL campaign of May–June 1987. They found Anomalous EIA inhibition and development simultaneously at widely separated American and Asian sectors accompanied respectively by events of morning electrojet reversal and evening partial ring current development. This

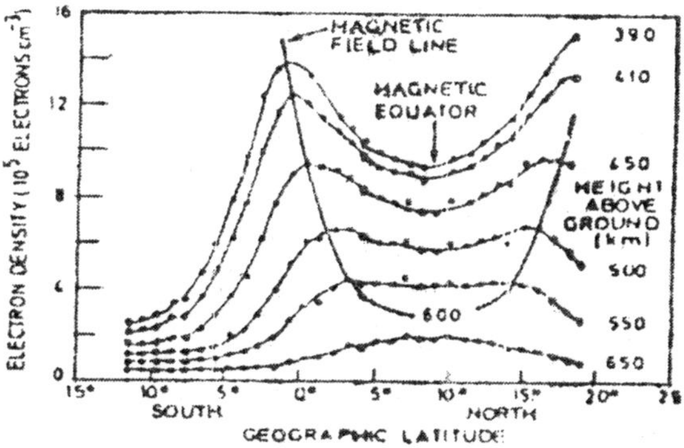

Figure 3: Latitude Variation of Electron Density (Electron Concentration) at fixed Heights Determined by means of the Aloutte I Top side sounder satellite above Singapore (1^0 N, 104^0 E) at 1234 local time, 15 September 1963. A magnetic field line is shown.

only shows that even if magnetic storm predictions are made with some reliability, it is not possible to predict the consequent changes in the ionospheric parameters in the EIA region.

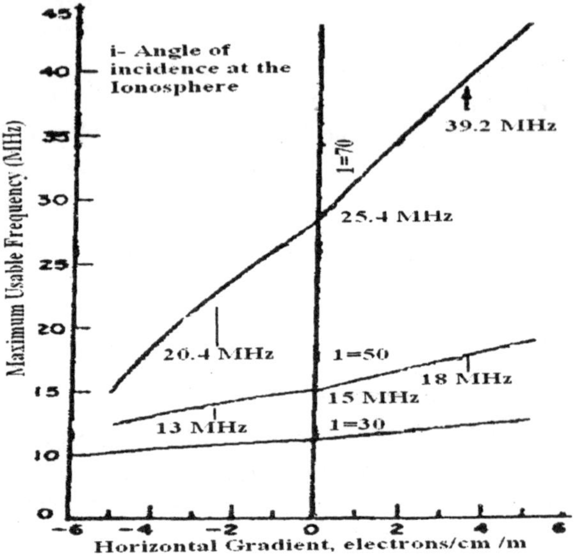

Figure 4: The Changes in MUFs Caused by Varying Values of Horizontal Gradients for Different Angles of Incidence

Fortunately, ionospheric storms do not create serious problems for HF communications in the low latitudes as already mentioned earlier in this section. The reasons are not far to seek. The heating of the high latitude ionosphere due to solar particles drives equator-ward winds. As neutrals are coupled to ions through collisions, neutrals will have to carry ions along. As ions tend to go parallel to magnetic field lines, these equator-ward thermospheric winds are propelled along magnetic field lines and hence move in the upward direction. When ionosphere (and the neutrals) is lifted up, to regions of lower recombination, f_oF_2 initially increases. But after several hours, as enough neutral molecular constituents accumulate in the F-region due to heating, the recombination rate drastically increases, resulting in a very significant collapse of foF_2 values at mid and high latitudes. But the nearly horizontal magnetic field in low latitudes is not conducive to lifting the thermosphere upwards and in any case, the force of the pressure gradient driving the winds gets weaker by the time they approach low latitudes. However, occasionally during night time the foF_2 values collapse near the equator (Figure 5) and may lead to communication problems. Ionospheric prediction and HF circuit availability are discussed in detail in several publications (See for example *Bradley and Bedford, 1976 and Reddy et al., 1979).*

(f) 'Fading of HF signals, especially in long distance systems, is to be studied not only to quantify margins in design, but also to understand propagation characteristics. The spread F irregularities mentioned in (c) above is only one of the causes. The

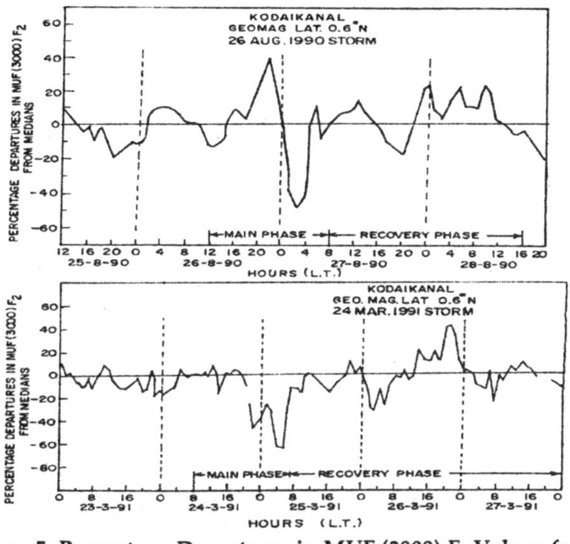

Figure 5: Percentage Departures in MUF (3000) F_2 Values from their Monthly Medians During Magnetic Storms of 26 August 1990 and 24 March 1991.

broadcasting requirements for field strength levels are higher by at least 20 dB above the requirements for radio telephony and proper margins will have to be given by monitoring real links for long durations. The ionosphere is inhomogeneous with continuous fluctuations and the field strength at the receiver is the net result of rays which have traveled from the transmitter via numerous slightly different sky wave paths. As a consequence, there is a random variation of the phase of each component with the resultant amplitude varying with time. Yet another reason for the signal fading is the variation of the polarization of the received signals due to the superposition of the ordinary and the extra ordinary rays which are differently polarized. As the polarization of the receiving antenna is fixed, polarization fading occurs. The dispersive character of the ionosphere makes the propagation parameters frequency dependent and this is sufficient to affect signal components within the communication band width and thereby causing distortion of the signal through selective fading. There are other bulk effects such as the slow variation of D-region absorption which produces slow fading. The fast fading approximates to Rayleigh distribution while the slow fading is more like the log-normal distribution. As HF links abound in low latitude countries, it is essential to study in detail these fading characteristics to quantify fade margins. Transmitter powers have to be increased to provide 10 to 13 dB margin in monthly medians or efficient (and expensive) diversity techniques will have to be used.

CONCLUDING REMARKS

The Internet-Cell phone revolution in the last 15 years has been phenomenal. To accommodate expanding requirements Sat. com. is going up to frequencies where rain and water vapour attenuation will set the limits for geographical coverage and data transmission rates. Apart from the fact that satellite transponder powers are at a premium, interference potential will also restrict the power. Thus optimisation of power with service requirements must take into account the various restraints imposed by the propagation medium. The availability of latest Sat. com. technology has not made such conventional modes of HF communications obsolete. HF systems continue to flourish all over the world but especially so in low latitudes. This thesis is an attempt to discuss the various problems, supplemented by data collected by the author and to provide the information on atmospheric constraints to the link designers. The results obtained through scientific analysis can provide inputs to ITU documents which are based on meager data sets from low latitudes.

CHAPTER–2

WIRELESS COMMUNICATION NETWORKS IN GOVERNMENT ORGANISATIONS IN INDIA

INTRODUCTION

The last fifty years have seen a revolution in the area of communications, essentially reducing the world into a global village. This Information Technology (I.T) has erased the boundaries between nations, continents, religions and between the rich and the poor. The cell phone culture has engulfed even the economically backward countries. Even the economy of several nation states are dominated by the I.T revolution. Several governments are exploiting these advancements in effective governance of their various wings. In this chapter, an attempt is made to describe some of the communication networks and equipment deployed in India by state and central government organizations, with special reference to the primary users such as Defence services, Police, Railways, Transport, Education and Health.

The author of this thesis has been involved for several years now in planning, modernizing and choosing the optimum combination of the networking described below:

1. HF Communication Systems.
2. VHF/UHF Communication Systems.
3. Satellite Communication Systems.

4. GPS Systems.
5. Cellular Communication Systems.

HF COMMUNICATION SYSTEMS

Need

For long distance point to point and point to multi point communications, Radio frequency band between 3 and 30 MHz, called High Frequency (HF) or "shortwave", is widely used since the invention of radio by *Marconi and Popov.* HF communications are growing at the moment despite the all-around implementation of cellular and satellite communication systems. HF communications is the only way of achieving global communication coverage without using expensive terrestrial and satellite infrastructure. HF communication still

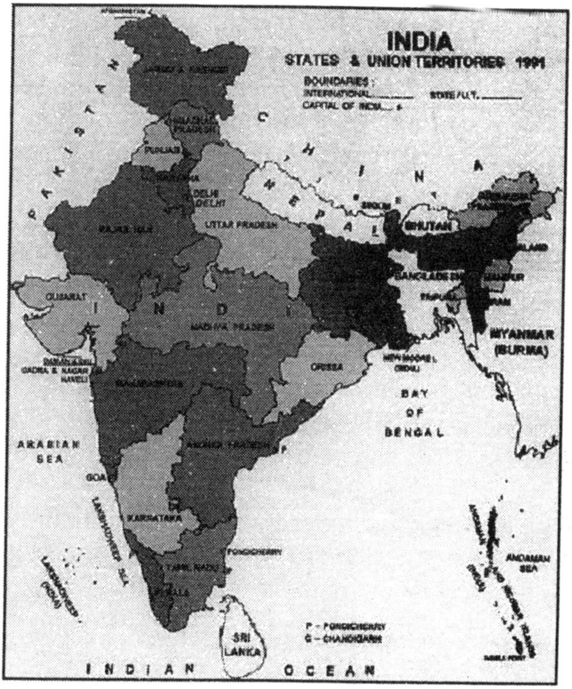

Figure 1: National Level HF (Morse code) Network for Long Distance Communication in India.

remains fail-safe means for communicating in low latitudes, including communication with mobile objects, e.g. for tracking airplanes on their routes etc., It is failsafe as in many instances, the user may have no control on satellite transponders.

The Primary Users of HF Communications

Marine & Fishing: Marine and fishing vessels require HF communications to reach thousands of miles and when a ship is far from shore, no other Communication systems would be as effective and inexpensive.

Aviation: Although aircrafts are often flying in areas where line-of-sight communication (such as UHF and VHF) are useful, HF communications is necessary for long-distance coverage.

Military: Various branches of military are using HF communications systems for base and field communications. In fact, during the desert storm operation of early nineties (1991) of last century, HF at near vertical incidence was used even for short range communications successfully.

Government: HF systems are used exclusively for sensitive, long distance communications and in remote regions which are difficult to reach- such as deserts, dense forests and mountainous regions and for embassy communications – HF systems play a vital role.

Law Enforcement: For long distance communication and for remote, inaccessible areas, for inter communications with in the forces that are deployed for tracking down extremists groups operating in dense forests and remote areas.

Amateur Radio: Amateur operators (known as 'HAMs') are licensed hobbiests who communicate via two way radio using HF frequency band for world-wide contact with Radio HAMs.

Commercial: Commercial users of the HF system typically use the system to contact personnel in distant or remote areas where any other medium is either impossible or too expensive.

Operating Modes of HF Communications

Morse Code (Continuous Wave)

Morse code which consists of dits and dots is widely used.

All amateur radio operators who use the HF bands are required to send and receive Morse code at least five Words Per Minute (WPM). CW might appear to be out dated now, as present day computers can send and receive many types of digital communications reliably. Still Law enforcing agencies like military and police in India use Morse code for their internal communication because of its simplicity and cost effectiveness and a modest signal to noise ratio. The specifications of a HF transmitter & receiver is given in table 1. The CW signal is very efficient because only the pattern of the signal (not the audio on the signal) needs to be understood. As a result CW is most reliable form of communication for human operators.

Voice Communications

HF equipment can be used for Voice communications. The modulation techniques are AM or SSB. Amplitude Modulation (AM) Consists of a base carrier modulated upper and lower side bands. Very few amateur radio operators use AM two way stations. The modern transceivers provide SSB apart from

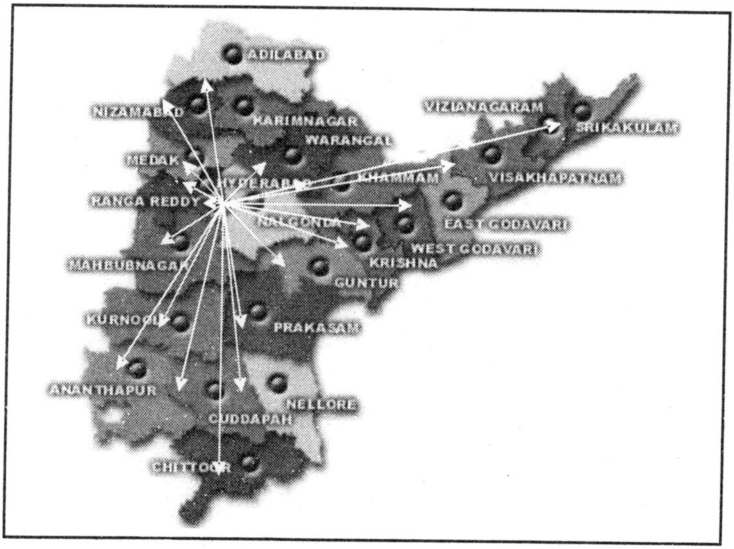

Figure 2: HF-voice Communication Network Covering all Districts of Andhra Pradesh

AM as single side band (SSB) technique is widely used for two way HF communications world wide as this is rarely hampered by fading and much more efficient than other voice modes. A typical HF voice communication network covering all districts of Andhra Pradesh, India is shown in figure 2.

Data Communications: RTTY

Radio Tele Type (RTTY) is one of the earliest forms of data transmission. In this system, printed data is transmitted via a high speed machine, rather than a hand-keyed, as is the case with CW communications. Although CW is a form of data communications, it is typed by a human operator. In order to send and receive RTTY, we need interface equipment. In addition to the standard HF transceiver and antenna combination, now the most common arrangement would be to interface a personal computer with the transceiver with appropriate demodulating/modulating software. The HF transceiver (Figure 3) normally used for long distance communication and produced by commercial manufacturers, has typical specifications as outlined in Table 1.

Table 1
HF Transmitter & Receiver Specifications

General Specifications	
Receiver frequency range	50 KHz. to 30 MHz.
Transmitter frequency range	1.8 MHz. to 30 MHz.
Emission types	AIA (CW), J2B, J3B (USB, or LSB)A2E (AM)
Frequency stability	+ 10 PPM from 0° to 40° C
Operating temperature	–10° to + 50° C
Antenna impedance	50 Ω nominal
Supply voltage	13 V DC ± 10 %
Power consumption	1.2 A (receive, no signal)20 A (transmit, 100 W output)
Transmitter	
Power out put	100 W (J2B, J3E, A1A)25 W (A3E)
Modulation type	SSB, AM
Spurious radiation	40 dB below peak out put
SSB carrier suppression	>40 dB below peak out put
Microphone impedance	500 – 600 Ω

Contd.

Receiver	
Carrier type	Double–conversion super heterodyne
Intermediate frequencies	Ist 47.055 MHz. 2nd 8.215 MHz.
Sensitivity (for 10 db S/N)	For 0.5 to 1.8 MHz., is 8 μV For 1.8 to 30 MHz., 0.25 μV
Squelch sensitivity	Better than 2 μV (1.8 to 29.999 MHz)
IF rejection	Better than 60 dB
Image rejection	Better than 70 dB
Selectivity	2.2 KHz. (-6 dB)5.0 KHz. (-60 dB)
Clarifier range	± 300 Hz.
Audio out put	At least 1.5 w into 4 Ω

Figure 3: A Typical HF Transceiver

VHF COMMUNICATIONS

Very High Frequency (VHF) range is from 30 to 300 MHz, and the signals propagate principally by Line-Of-Sight (LOS). The range of frequencies normally used are LB (Low Band) VHF from 60 to 80 MHz and HB (High Band) VHF from 130 to 180 MHz. The security agencies avoid using LB VHF frequency as these frequencies can be easily monitored using

FM band radio receivers. Although LOS restrictions limit the ground range of VHF systems, LOS is an effective means of ground communication for distances up to 50 Km (depending on terrain and antenna height, transmit power) without using a repeater. By placing repeaters properly along the required communications path, long-range VHF transmissions through a series of short LOS hops is possible. Depending on the use, range, and number of channels intended, VHF transmitter power can range from 2 watt for a portable hand-held FM radio to several hundred watts for a 12/24 multi channel LOS system. The specifications of a typical VHF trans-receiver is given in Table 2. The higher the frequency, the less power required to transmit VHF signals over a given distance. An example of covering a district Kurnool of Andhra Pradesh with two VHF repeaters is shown in Figure 4.

Single-channel VHF radios are portable, vehicular, or airframe mounted, and can usually be operated in motion. The larger multi channel systems are commonly mounted

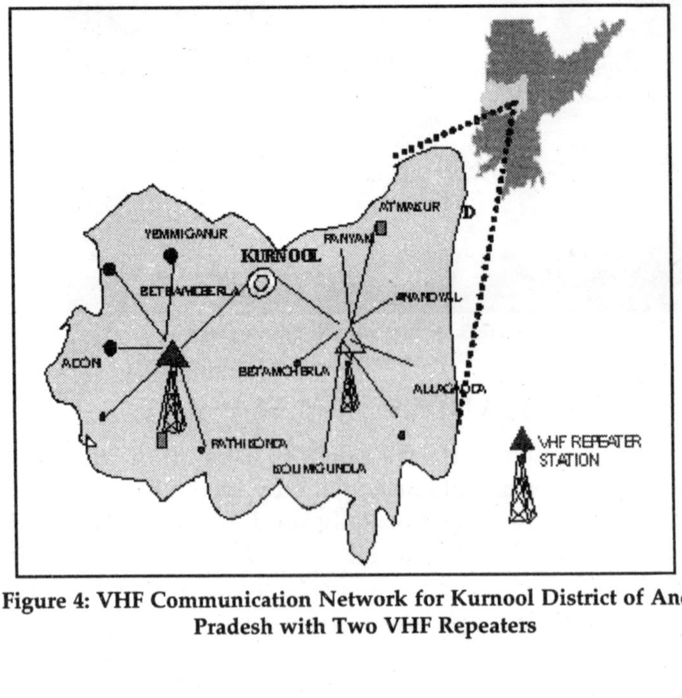

Figure 4: VHF Communication Network for Kurnool District of Andhra Pradesh with Two VHF Repeaters

aboard ships and trucks and requires careful orientation of directional antennas. Typical uses include short-range FM combat radio nets, radar, radio navigation, wideband LOS multi channel systems (repeated or non repeated) and television broadcasting. VHF links can provide excellent circuit quality, comparable to cable systems with up to 99 percent reliability. VHF links can handle either analog or digital transmissions in single/multi channel modes. Data rates may vary from 45.5 to 75 bits per second (bps) for mobile VHF radio nets to 1.2 to 9.6 Kbps per channel for LOS multi channel radio relay systems.

Table 2
VHF Transceiver Specifications: (Handheld Set)

General	
Frequency range	136 MHz to 174 MHz.
Channel Capacity	16 to 128 Channels
Power Supply	7.5 Volts ± 20%
Average Battery life	> 8 hrs
Transmitter	
RF Output	Low 1 W (or) High 5 W
Frequency	136 MHz. to 174 MHZ.
Channel Spacing	12.5/20/25 KHz.
Freq. Stability ((-30° C to + 60 ° C)	0.00025%
Spurs/Harmonics:	-36 dBm<1 GHz.-30 dBm >1 GHz.
Audio Response	+1, -3 dB
Audio Distortion:	< 5%
FM Noise	-40 dB
Receiver	
Frequency	136 MHz. to 174 MHz.
Sensitivity @12 dB SINAD:	0.35 μV
Adjacent Channel Selectivity	-70 dB
Inter modulation	-65 dB
Freq. Stability (-30° C to + 60 °C)	0.00025%
Spur rejection	-70 dB
Image Rejection	-70 dB
Audio Out put @ < 5% distortion	500 mW

Ultra High Frequency (UHF) Communications

The UHF range is from 300 to 3,000 MHz. The main propagation methods include tropospheric scatter, satellite, air/ground and LOS. Due to the flexibility of UHF communications, the distance range varies significantly. LOS coverage is 20 to 140 Km, depending upon the terrain. In this type of communication systems, transmitter power can range from a low of 10 to 100 watts for LOS and satellite systems, while troposcatter systems operate in the 2,500 to 10,000-watt range. Many UHF systems are transportable by vehicle, aircraft or ship. Some UHF/satellite terminals are small and light enough to be portable. Common UHF applications are seen daily in local ambulance, fire and police radio nets, with repeater operations being typical. On military installations, the non-tactical Intra Base Radio (IBR) nets are usually VHF or UHF. UHF systems are capable of high-quality, reliable and high-capacity transmissions with data rates of 2.4 Kbps and higher. UHF is used widely to provide secure/non secure voice, data and facsimile services in both mobile and fixed configurations. Along with VHF, UHF is the band preferred for television.

Some typical Frequency Bands in use are: 400- 800 MHz. Trans- receivers of 25 W Transmit power and receiver sensitivity of 0.25 micro volts are used for static, vehicular & mobile operations which can provide range up to 30 Km without repeater. Hand held sets of 5 W transmit power are being used which can provide a range up to 5 Km. The range of Communication can be extended using one and two repeaters. This concept is useful for establishing VHF/UHF communication link for longer distance. Main users are Police, Municipal services, Health, Electricity and Fire Services.

The VHF/UHF communications are reliable throughout day and night, unlike HF communications. Data Communications links are also established using VHF networks in inaccessible areas. In simplex mode, for two way communications one RF channel is used. Different channels can be used for different group communications. Trans-receivers of 64 channels capacity are commonly available for accessing different channels. Digital sets up to 255 channels are also available. Communication secrecy can be installed using encryption devices called scramblers in VHF/UHF sets for speech secrecy. Ten million code combinations are programmable in these secrecy devices.

Radio Trunking Systems

These networks operate in VHF/UHF frequency bands 150 to 200, 400 & 800 MHz. The main advantage is to ease traffic congestion when more number of users are trying to access one channel at any point of time. Here more than two channels can be made available for access. Even though the system is costlier, it provides broadcast call which is most important for security agencies. This network can also handle voice, data & video by incorporating protocols like APCO-25 and TERTA. The system selects a free channel automatically and re-allots to other user. This has connectivity (optional) to the PSTN, for telephone access from any mobile location as shown in Figure 5. Radio Trunking industry standards are based on Common Air interface/code word signaling protocols. Several protocol systems were developed over many years. LTR (Logic Trunked Radio) system is developed during the year 1970. MPT (British Ministry of Posts & Telegraphy) 1327 &

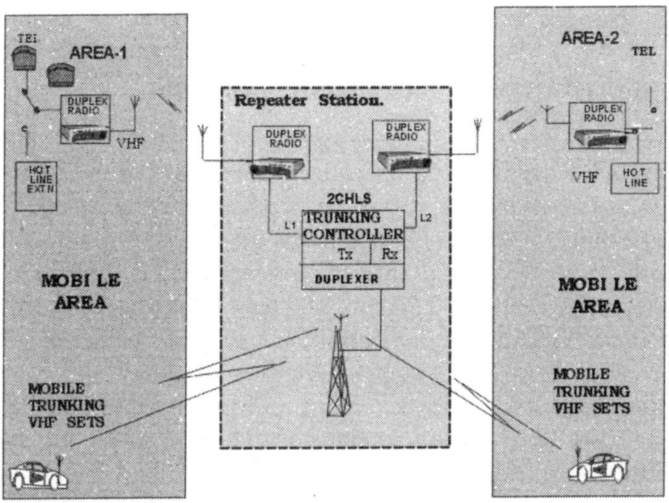

Figure 5: Radio Trunking System Connecting Two Metro Areas

1343 are developed during the year 1984. APCO (Associated Public safety communications officers) APCO-16 is developed during the year 1985. APCO-25 is developed around 1994. TETRA (Trans European Trunked Radio), digital TDMA protocol system is developed recently to cater to high speed data and video. Main users are Industry sectors for distribution & delivery. Financial sectors like Banking, stock brokers, construction companies, Medical, Transport, Electricity etc.

Table 3
Specifications of Inmarsat Phone

Frequency: Transmit
1626.5 – 1660.5 MHz
Receive
1525.0 – 1559.0 MHz.,
Mode of operation
BPSK – QPSK – FDMA
Receiver sensitivity
139 dBm
Power consumption
40 W (S/Bi)/120 W mode
Input supply
AC and 10-35 VDC
Battery Pack
Ni Cd 5 AH/18 V

SATELLITE COMMUNICATION SYSTEMS

There are over 750 satellites, many of which are primarily available for use in communications. The orbit of the satellite either polar, meaning it travels around the earth over the poles, or geo-synchronous, where the position is fixed (mostly) somewhere over the equator. Geo-synchronous orbits are at 36,000 km in altitude. The time delay for the signal to travel to the satellite and back is about 0.3 sec and is noticeable. Polar orbits are much lower, so there is no delay. Polar orbit satellites travel around the earth any where from 1-12 hrs, and require many satellites for complete World coverage. Satellite communication frequency bands must be UHF and higher in order to penetrate the ionosphere, and are commonly in the SHF range. They are generally either C band (3.75-7.5

GHz), Ku band (12-18 GHz) or Ka band (27 to 40 GHz) (Ref. Figure 6). Further more, circular polarization is predominately used since it is difficult to properly orient the receiving antenna to match the satellite antenna if linear polarization is used.

INMARSAT Satellite Communication Equipment

INMARSAT is an international organization, set up in 1979 to provide world-wide mobile satellite communications, for commercial, emergency and safety applications on land, at sea and in the air. Recently INMARSAT-3 has enabled the introduction of satellite phones smaller than Laptop Computer, with advanced features found in cellular systems. INMARSAT satellites provide voice, fax and data communication services.

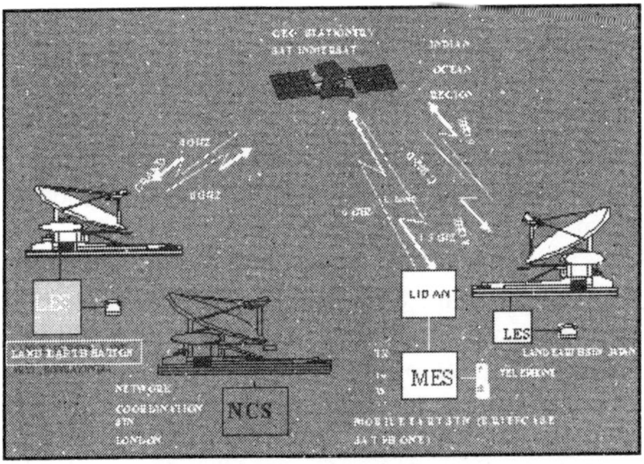

Figure 6: Inmarsat Satellite Phone

The INMARSAT satellites lie in geo-stationary orbit 36,000 km distance in space. Each satellite covers up to one third of the Earth's surface and is strategically positioned above one of the four ocean regions for continuous global coverage. It will use five high-powered spot beams and one global beam to serve the four Ocean regions to cover more than 95 per cent of the globe land mass. The INMARSAT phone is a brief case version. The upper cover houses the antenna and the lower houses the equipment. The brief case AZ and EL angle to be set in such a way to see the Indian ocean Satellite for India and to be adjusted for maximum received signal for clear communication.

Indian Satellite Communications Systems –INSAT

INSAT (Indian National Satellite System) series of satellites are multipurpose satellites for telecommunications, broadcasting, meteorology and search & rescue services. Commissioned in 1983, INSAT is the largest domestic satellite communication system in the Asia-Pacific Region. Indian Space Research Organisation (ISRO) is Indian national space agency. ISRO was setup in 1969 under India's Department of Atomic Energy. Its first satellite was launched in 1975 by a Soviet booster and its first domestic launch in 1980. India created the space commission and Department of Space in June 1972.

Satellites in Service

INSAT-2C: Launched on December 7, 1995 is located at 93.5^0 E longitude.

INSAT-2DT: To partly augment the capacity of the INSAT system consequent to the loss of INSAT-2D (in October, 1997), is located at 55^0 E..

INSAT-2E: Launched on April 03, 1999. It is located at 83^0 E longitude. Under an agreement with the International Telecommunications Satellite organisation (INTELSAT). Eleven 36 MHz equivalent units of C-band capacity on INSAT-2E have been leased to INTELSAT under a commercial agreement.

INSAT-3B: Launched on March 22, 2000. It is located along with INSAT-2E at 83^0 E longitude.

INSAT Services

1. Telecommunications
2. Mobile satellite Services- Mobile Telephony & Message Reporting System
3. Television coverage
4. Radio Networking

5. Telemedicine
6. Meteorology
7. Satellite Aided search, Navigation & Rescue

INSAT Satellite-3B Ku Band Operation Description

INSAT Master Control Facility (MCF) at Hassan in Karnataka is responsible for initial and in-orbit operation of all INSAT satellites. It has a network of six Satellite Control Earth Stations (SCES) with a full motion 11 m diameter antenna and two 7.2 m diameter, limited motion antennas, commissioned to meet the requirement of INSAT-3 series of satellites. MCF has an advanced computer network with six servers and fifty client workstations exclusively dedicated to spacecraft mission operations. MCF also has an elaborate backup power system to ensure uninterrupted mission operations.

INSAT-3B will boost the Very Small Aperture Terminal (VSAT) services that have seen a rapid growth. Today about 45,000 VSATs are serving. The major VSAT users include banking and financial institutions, stock markets, goods sector and medium to heavy engineering companies. Several public and private sector units have established dedicated closed-user group networks for their in-house applications. INSAT-3B will almost double the transponder capacity for these services. For the first-time Ku-band frequencies will also be used for VSAT services, which enables the use of smaller ground terminals.

Very Small Aperture Terminals (VSATs)

As terminal technology advances, the size of the antenna required to achieve a particular link quality (bit error rate) decreases. A class of terminals smaller than VSATs is now available; these are termed Ultra Small Aperture Terminals (USATs). For most practical purposes, USATs are just VSATs with smaller antennas. It must always be remembered, however, that as antenna size decreases, the antenna beam widens and that a point is rapidly reached when there is no further advantage in decreasing antenna size because of increased interference with other systems. The practical present day lower limit on antenna size is 55 cm diameter.

The most common VSAT configuration is the TDM/TDMA star network. These have a high bit rate outbound carrier (TDM) from the hub to the remote earth stations, and one or more low or medium bit rate Time Division Multiple Access (TDMA) inbound carriers. With its star configuration network architecture, interactive VSAT technology is appropriate for any organisation with centralised management and data processing. This configuration has been developed to minimise overall lifetime costs for the complete network including satellite transmission costs. The use of a single high performance hub allows the use of low cost remote VSAT terminals and optimises the use of satellite capacity. Even so, in most VSAT networks, the cost of the VSAT terminals

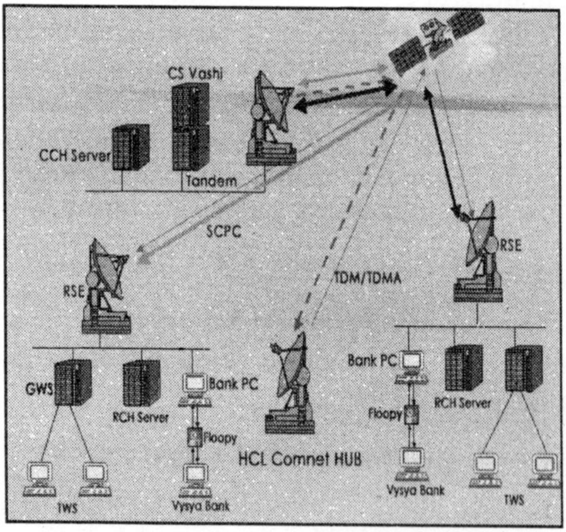

Figure 7: VSAT-Hub Network of Stock Exchange

usually far exceeds the cost of the hub (typically a VSAT terminal is 0.1 to 0.2% of the price of the hub). In a typical VSAT network as shown in Figure 7, remote user sites have a number of personal computers, dumb terminals and printers connected to the VSAT terminal which connects them to a centralised host computer either at the organisation's head office or data processing centre. Data sent to the VSAT terminal from the DTEs (Data Transmitting Equipment) is buffered and transmitted to the hub in packets. Communication Satellites provides an efficient method of sending Radio, Telephone, Telegraph signals and other data over a vast area. The development of Satellite Communication technology has not only improved the international communications but also domestic communication system. This has been demonstrated by INSAT–for domestic Communications in India. VSAT networks have the facility to add new sites and to change the capacity of the network. VSAT, networks through satellites can provide Wide Area Networks (WAN) by interconnectivity of LAN and other networks. In India some of the VSATs network operators are NIC (National Informatics Centre), Banking sectors, Stock exchanges etc. as shown in Figure 7.

GPS–GLOBAL POSITIONING SYSTEMS

The Global Positioning System is based on 24 earth-orbiting satellites. Each transmits a very precise timing signal from an on-board atomic clock. A GPS receiver anywhere in the world can receive these signals and from minute differences in their timings figure out the exact distance to each satellite. Using simple geometry, it thus knows exactly where it is on the planet. Designed by the United States Department of Defense for use by the Military, the GPS was made available to civilians for use in aviation, marine, surveying, general out door recreation markets. Today millions of people utilize the accuracy and convenience of this amazing technology to hunt, fish, fly, Cruise. GPS co-ordinate systems are positions or locations that are described in a unique way. Different coordinate systems are LAT/LON, UTM, TD, OSGB, Irish Grid, Swedish Grid, Swiss Grid, Fennish Grid. UTM (Universal Transverse Mercator) and LAT/LON are most commonly used co-ordinate systems.

Satellites transmit two codes. PPS - Military only encrypted code and SPS (Standard Positioning Service) codes. All GPS receivers are SPS receivers. Satellite transmits two types of data–Almanac & Ephemeris. Almanac is the general information on the location and health of each satellite and Ephemeris is the precise satellite positioning information used for ranging. Both almanac and ephemeris data is used for a GPS receiver to locate and acquire satellites quickly and compare a position fix. DGPS is a Differential GPS which compares the size of the error and applies it to positioning information.

Figure 8: Deployment of GPS Receivers for Specific Operations

As ionosphere is a dispersive medium, it is possible to use two or more GPS frequencies and make correction for the ionospheric group delay which is caused by the total number of electrons in the ionosphere. But single frequency systems need an average Total Electron Content (TEC) input to make the correction. Models developed for use in mid-latitude countries are not accurate for India because of EIA gradients and large diurnal variation in TEC. *Wintoft and Cander (2000)* used time delay neural networks with back propagation for hourly $f_o F_2$ predictions with some success. *Sarma et al., (2000)* described reference station placement for the Indian Wide Area Augmentation System (INWAAS) and it is expected that adequate data from India will be generated soon to correct for ionospheric and tropospheric errors in GPS.

CELLULAR COMMUNICATION SYSTEMS

Cellular Networks

The cellular network uses 800-900, 1100-1200 MHz frequency bands to connect mobile stations (Cell Phones) to central receiving antennae and base station. The area of coverage by that base station defines the cell. The base stations are interconnected by fixed media similar to the telephone network. A central system takes care of locating the recipient and the signal is sent to the appropriate antenna in that cell for broadcast, or alternatively passed on to the fixed telephone network.

When a cellular subscriber moves from one cell into another, the base stations perform a handoff and transfer the call to the next base station. Even when not placing calls, cellular phones send low power signals in order for the system to determine the phone's location for incoming calls. Somewhere around 50 MHz of bandwidth is available divided between the transmit and receive side. Most cellular phone signals are MSK (Minimum Shift Keying, a variety of FSK) and have a signal bandwidth of 200 KHz. That gives a capacity of about 125 channels per cell.

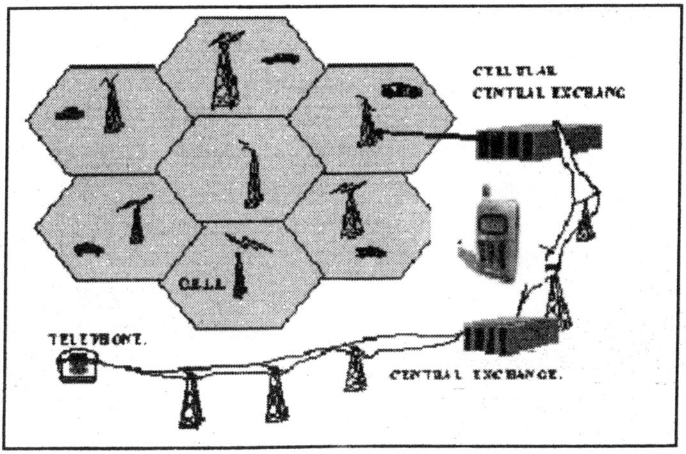

Figure 9: Cellular Network Scenario

A cellular network allows cellular subscribers to wander anywhere in the country or any part of the World and remain connected to the *Public Switched Telephone Network (PSTN)* via their mobile phones as shown in figure 9. A cellular network has a hierarchical structure and it is formed by connecting the major components mentioned in the figure 10. All BSCs within a cluster are connected to a Mobile Switching Centre (MSC) using land lines. Each MSC of a cluster is then connected to the MSC of other clusters and a PSTN main switching centre. The MSC stores information about the subscribers located within the cluster and is responsible for directing calls to them. The size of the cell largely depends on the area in which the cell is located. Generally, rural areas have less subscribers compared to urban areas. So in an urban area more channels are needed to accommodate the larger number of subscribers. If each cell in a given rural and urban area had fixed number of channels, the cell size in the urban area would have to be smaller to allow more channels in the given area. Reducing the cell size would result in cells, using similar channel frequency, to be located closer to each other. Therefore reducing the size too much would cause an increase in co-channel interference. Size of the cell can be varied by varying the power and sensitivity of the base station

A cellular network could also be used to transmit *data, images, text and video.* A lot of research has been done and is still in progress to maximize the ability to transmit these other forms of information via cellular networks which are commercially available. All these new inventions would require the existing cellular network to be improved. One of the most important improvement is to create a single global standard for cellular network to transmit all forms of information. This would allow information to be transmitted throughout the world and eventually make the world smaller and more united.

FUTURE PLANS-MODERNISATION OF COMMUNICATION SYSTEMS

Satellite Communications-Hand held Devices

Thuraya Satellites launched in 2001 is a regional mobile communication system that provides satellite telephony and data services to Europe, The Mediterranean, Northern Africa, the Middle East and other Asian regions (covering more than 100 countries) as shown in Figure 10. Thuraya Satellite Telecommunications Company established in April 1997, is a consortium of over 21 national telcos and investment houses. Utilizing two geostationary satellites and a ground station based in Sharjah, UAE, Thuraya offers a unique product mix

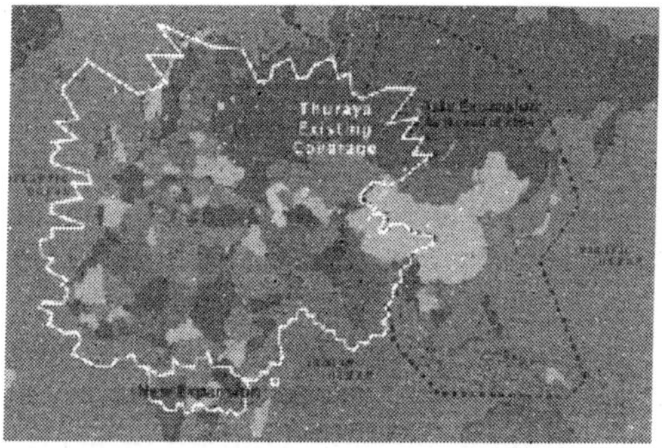

Figure 10: Thuraya Global Coverage

Figure 11: Thuraya Satellite Handheld Set

of mobile satellite and GSM services from one small handheld satellite phone as shown in Figure 11. Thuraya has roaming agreements with many global phone GSM operators around the world which allows ease of use when satellite coverage is not required (within populated regions). Thuraya was the first mobile satellite service to truly integrate satellite and terrestrial mobile services through one handset, allowing users to move freely in and out of terrestrial coverage or within structures which impede satellite coverage.

GIS, GPS & AVLS Networks

The US Geological Survey defines GIS or Geographic Information System as 'a computer system capable of assembling, storing, manipulating and displaying geographically–referenced information, that is, data identified according to their locations'. This means that different types of data (residing on a database) can be presented as layers, one on top of another, and then placed on a digitized map. The advantage of GIS is that it can visualize and can show a 3 D scenario of a process or action in a few seconds or minutes that could otherwise take up a few weeks if done manually.

Geographical Information System (GIS)

GIS is a boon for military commanders who are often deployed with in a short notice to different battlefield scenarios. One of the essential skills needed to be the overall in charge of Military or Police operations is in obtaining information about the Territory topography and relief features from detailed maps and then formulating plans for troop movement. Traditionally, sand and mud relief models have been used to simulate possible offensive or defensive actions. GIS models can help the commander brief his 'boys' about the topography, possible routes for infiltration, rescue of hostages and extraction of the special team after the operation. He can also simulate possible courses of action that the operation might take. For example, the commander could use it to simulate the effect of rain and slush and the subsequent delay these factors would have caused in an operation. Using a GIS model, it is possible to predict how much the roads should be widened, what kind of drainage facilities should be provided, and how many buildings and people should be relocated to facilitate smooth flowing of traffic. A GIS system is made up of the required hardware, GIS software and a database. GIS is much more than digitized cartography. GIS can also be combined with GPS receivers to simulate or study situations in emergencies or natural disasters. An example of using GPS, GIS, AVLS (Automatic Vehicle Location System), & Digital Radio Trunking system for the deployment of patrolling vehicles for enforcing law and order is shown in Figure 12.This system is developed integrating GIS and GPS Technology. This will track vehicles as per predetermined algorithms and almanac and determine the position of the vehicle to an accuracy of about 50 meters or less with additional infrastructure. With differential correction, it is possible to achieve accuracy less than a meter. AVLS is fully network compatible, can work via internet, can automatically recognize all vehicles in motion. Primary users are Fleet movers, Security Agencies, Police and Fire Services.

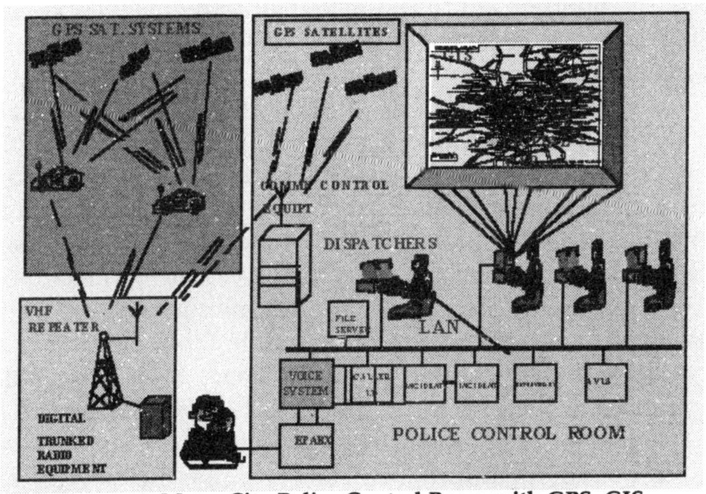

Figure 12: Metro City Police Control Room with GPS, GIS, AVLS & Digital Radio Trunknet Systems

CONCLUSION

Convergence of telecommunications and computers is the important implication of information technology. This convergence has a major impact on technology, economy and society. Whereas computer provides-precision and storage of information, it is the telecommunication that provide a connected communications world wide and global communication enhances social and economic growth. All Government organizations are continuously modernising and improving their communications infrastructure and capabilities to serve the Society effectively. However, a coherencive and exhaustive study on the propagation impairments on these frequencies is very much required to provide optimum service to user community.

Availability of service should be ensured for 99% of time. In this context it is very much required to study the atmospheric constraints for HF, VHF, UHF and higher frequencies. The aim of this study is to identify suitable models for the Radio wave propagation in the above bands. The study also compares the adaptability of ITU-R models in general to tropical countries and in particular to Indian sub-continent. The results are expected to be useful for future planning of radio networks in India.

As plans are on the anvil in India to use the Ku band (12 to 18 GHz), Rain morphology and Rain attenuation become very critical. At the upper end of Ku band, even water vapour attenuation is important. As there are hundreds of HF links operating in India, fading in HF, Atmospheric Radio Noise, HF frequency predictions, etc., are of concern. Some of these problems are addressed in the following chapters with observational data wherever possible from a number of stations in India. In the UHF, the primary focus for cellular operations & Radio trunking, is the 800 to 900 MHz band. It is essential to study propagation characteristics/path loss etc., in this band for urban/semi-urban areas in view of the extreme demand in cellular traffic.

CHAPTER–3

ATMOSPHERIC RADIO NOISE MEASUREMENTS IN INDIA-AT LOW LATITUDES

INTRODUCTION

Atmospheric Radio Noise (ARN) is a crucial parameter in limiting the efficiency of Radio systems operating in HF range and it is in this context planned measurements of radio noise were made from 1957 till 1961 at sixteen different stations in the world including India. All India Radio (AIR) had conducted measurements at four stations in India representing various regions of the country *(Ghosh and Mitra,1959; Ghosh et al., 1974)*. From these measurements it was observed that those predicted from *CCIR report (1963)* are significantly at variance with the measurements conducted at various places in India. National Physical Laboratory has been carrying out measurements of average level of ARN since 1987, using standard field strength meter equipped with vertical rod antenna. Results of this systematic study of ARN levels carried at NPL and at Tirupati during 1997-98, vis-à-vis earlier related studies are presented in this chapter. The NPL and Tirupati measured values are generally higher on all the frequencies, when compared *with CCIR (presently ITU-R)* predicted values from the *report 322 (CCIR, 1985)*. NPL, AIR and Tirupati measured values do not exhibit the dramatic increase in noise levels during the transition period (1600-2000 hrs IST) in contrast to the CCIR predicted values. This study has shown that there are certain major disparities between the actual ARN observations and CCIR predictions for India. This study has also shown the necessity of extensive measurement of radio noise in different regions of India, especially during the transition period.

ATMOSPHERIC RADIO NOISE

Noise is any undesired form of energy interfering with the proper and easy reception and reproduction of wanted signals. In electrical sense it is called radio noise. In any radio communication system, noise sets a limit for the satisfactory reception of radio signals. Radio noise can be classified in two major groups, external and internal noise. External noise is constituted of atmospheric radio noise (ARN), extra terrestrial noise like solar, cosmic or galactic noise and man made noise due to electrical appliances etc., internal noise includes thermal agitation noise, shot noise and transit-time noise. External noise can be reduced only by changing the location of receiving equipment, whereas internal noise can be brought down by suitable design of equipment.

Atmospheric Radio Noise (ARN) also called static is caused by the naturally occurring inter cloud electrical discharges in thunderstorms accompanying electric flashes in the earth's atmosphere. They are spread over a wide range of radio spectrum and their intensity bears an inverse ratio to the frequency. Their effects are significant up to 30 MHz which includes the broadcasting bands. Galactic and man-made noise also have adverse effects in the HF bands. *CCIR Reports No.322 (1963-1985)* give characteristics of radio noise up to 30 MHz. This noise is caused by local and distant thunderstorms. It has been shown by *Winn et al., (1974)* that in tropical zone, 90% of electric flashes are due to inter cloud electrical discharges taking place at a height of 5 to 8 km with an average height of 6 km above the mean sea level. Accordingly the horizon for these noise bursts may be around 250 to 300 km. It is stated that the height of cloud may come down even to height of 3 km in the tropics thus reducing the extent of horizon. A local thunderstorm is defined as one whose radiation can be directly received without reflection from the ionosphere. This kind of thunderstorm occurs with in an area covered by a radius of 500 km or even less. The noise from the local thunderstorm is of high amplitude. In the tropics the local thunderstorm days exceed 50% of days of the season. This activity begins around noon and continues till the following sunrise. Apart from impulsive noise, fluctuating type of noise is usually caused by the distant thunderstorms and distant noise is propagated via the ionosphere. However, the contribution of noise due to distant thunderstorms is significant during nights only. In general it has been observed that the noise levels are enhanced during 1600-2000 hrs. The ARN power at a certain location is the sum of local and propagated components of noise.

Atmospheric Radio Noise in the HF band decreases rapidly with increasing frequency. The two components of ARN have distinct morphologies. One component is due to lightning discharges occurring close to the receiving site and the radio noise is propagated directly from the point of discharge to the receiving antenna. The range of this mode of propagation depends on the altitude of the active clouds and the cloud altitudes are known to be higher in the tropics. Consequently in the tropics, clouds in a larger area around the receiver contribute to this component of ARN. So, technically the LOS range modified by the troposheric refraction (the modified Earth Radius concept) decides the radius of the circle from where ARN reaches the receiving antenna. The second component is that due to cloud discharges beyond the zone from where the radio noise travels to the receiving site via the ionosphere. It is this segment which exhibits large diurnal and frequency variations. Firstly, the non-deviative absorption in the D-region of the ionosphere at the lower end of HF band is prohibitively large and very little ARN power will reach the receiving site at lower HF spectrum. This is particularly true in the day time; but at night, D-region disappears and the non-deviative absorption diminishes. Thus, it is clear that ARN undergoes both local time and frequency variation. This is to be coupled with the fact that the production of ARN is highly frequency dependant and decreases rapidly with increasing frequency. The CCIR values are biased with data from mid latitudes where the cloud heights are lower, resulting in the smaller component from the LOS type of ARN.

There are certain problems related to ARN that are critical only at equatorial and low latitudes. Most of the countries in these latitudes are developing countries which depend on inexpensive HF links for a variety of operation in civil, defense and para-military services. Not only the generation of ARN is high at low latitudes, but also the propagation mechanisms favor preponderance of ARN even up to relatively high frequencies. One such mechanism is sporadic E present most of the time around the equator, especially during day time. This sporadic layer promotes propagation ARN from long distance cloud discharges with very little attenuation. Some experiments conducted at Trivendrum (over the magnetic equator) clearly showed that ARN values were very high even at 7 MHz and the correlation with presence of sporadic E layer was very good *(Ghosh, et al., 1974; 1978)* .

Yet another reason for the critical importance of ARN at equatorial and low latitudes stems from the fact that the F region collapses rapidly to very low values in this region (obviously due to the low electron and thermal content of low latitude geomagnetic field tubes), forcing HF link operators to opt for lower frequencies where the ARN is prohibitively high. In some areas, during times of high thunderstorm activity, ARN could be the limiting factor reducing service availability to unacceptable levels for many services. It is instructive here to take a comprehensive look at the data available from other sources and compare with our own measurements comprehensively *{Kotaki et al., (1981)}*.

As already stated, *CCIR report 322 (1963)* gives the characteristics of atmospheric radio noise. Large differences have been observed by many workers *(Herman, 1961; Ibukin, 1965; Joglekar,1971; Ghosh and Saksena, 1978)* between the measured values and the values predicted from the *CCIR Report*

322 (1963) especially in the tropics. In this context CCIR formulated a study programme (29 B/6) and among other things this study programme aims at the study of world distribution of noise power with existing or augmented network of stations. Accordingly National Physical Laboratory (NPL), New Delhi under took the measurement of ARN levels during 1987-88 at selected frequencies of 2.5, 5, 7 and 9 MHz and at Tirupati during 1997-1998 at 2.5, 5.0, and 10.0 MHz. In this chapter a comparative study between the measured and predicted values of ARN levels using different methods has been made.

A BRIEF REVIEW OF PAST WORK

ARN level measurements in HF bands were started in U.K, USA, India and Japan in the thirties, but inter comparisons of results of these measurements were not possible since the methods of measurement of noise were not standardized. Around 1950's, in view of increasing demand for allocation of frequencies for broadcasting and other communication systems, a need was felt to establish a network of noise measuring stations using standardized method of measurement. As per the CCIR recommendation in 1956, planned measurement of ARN were made at sixteen stations all over the world, including India during the years 1957-61 and a report on "World Distribution of Characteristics of Atmospheric Radio Noise" was presented by the CCIR in 1963. However, since only sixteen stations over the whole world were involved, the report could not be comprehensive. All India Radio measured ARN levels at four stations in India namely New Delhi, Gauhati, Vizag and Trivendrum during 1957-1965 and their results were presented in *CCIR Document No.303-1 (1966)*. *Aiya (1977)* in a comprehensive paper discussed the topic of atmospheric noise as a source of interference to analog and digital communication and also reviewed the available information on thunderstorm, thundercloud and lightning. He has provided new noise level data from 0.3 to 30 MHz for tropical zone using a radio noise meter developed by him. He suggested the need for taking measurement every hour rather than in 4 hours block. *Joglekar et al., (1975)* from the analysis of noise data of the *CCIR report 322 (1963)* in the time block 2000-2400 hrs have observed a decrease in noise level with increasing sunspot activity at 5 MHz due to residual absorption. Above 5 MHz there is a tendency to increase due to favorable propagation conditions. *Huang (1977)* from the analysis of noise data collected at Taiwan (1967-72) did not observe any change in the ARN level in the frequency range 2.5–20 MHz. At 5 MHz, he did notice a slight negative dependence near noon. Huang also observed some difference of noise levels tending to be greatest near noon as compared with *CCIR report 322* .

Agrawal (1983a) from the analysis of noise data (median and upper decile values) collected at New Delhi on 2.5 and 5 MHz using an ARN-2 recorder, observed systematic variations of noise level in the evening transition period. In another paper *Agrawal (1983b)* studied the frequency characteristics from 1-10 MHz and observed variations of ± 6 dB in the time block 1600-2000. *Agrawal (1983c)* has found out empirically a conversion ratio of about 10 dB (with 4 dB standard deviation) between Aiya's method and that of NBS (ARN-2 recorder) method.

Kotaki et al.,(1981) have derived maps of global distribution of thunderstorm activity observed with Ionospheric Sounding Satellite (ISS-b), from the data collected during 1978-80. Recently global maps of occurrence of lightning discharges have also been made by *Kotaki and Katoh (1983)* and based on information on lightning discharges they have further derived global maps on atmospheric radio noise for 2.5, 5, 10 and 20 MHz on hourly basis for autumn season. The radio noise values derived by them were found to be in better agreement with the observations made all over the world when compared to data from *CCIR report 322 (Kotaki, 1984)* .

Sailors and Brown (1983) have developed a microcomputer based model of ARN in UT for 1 MHz median noise level. In the worst season (summer), the rms error of the most accurate model was 1.4 dB whereas the rms from the *CCIR Report 322* was found to be 6.7 dB.

Kotaki and Katoh (1983) have observed that in the lower part of HF band (2.5-5 MHz), the local component of noise is very large around 99% during day whereas the local component in the nighttime is hardly 1% and the propagated component of noise is very large at night. As the frequency increases upto 10 MHz the day time local component becomes around 88% and its nighttime value is only 10%. The propagated component is largest at night say around 90%. As the frequency increases upto 20 MHz, both the components of the atmospheric radio noise become almost equal during night time. It has been observed that at the higher end of the HF band, the difference in the amplitude distribution of ARN in the daytime and night time is significant. This fact is borne out even by the experimental observations carried out at NPL, New Delhi, during 1987-88. Form the knowledge of propagation losses at different frequencies in the HF band, it has been observed that during daytime the amplitude spectrum of noise depends upon the amplitude spectrum of local thunderstorms. However, during nighttime proportion of the propagated component of ARN is large, and as this component is originated by many sources located far off in the world, it is difficult to explain the dependence of ARN quantitatively. A precise knowledge of amplitude probability distribution of noise at a location will help in the efficient design of telecommunication system.

DATA BASE

The main aim here is to draw comparisons between the measured values of average ARN levels by All India Radio, CCIR predicted values and those estimated by *Kotaki and Katoh (1983)* from thunderstorm activity data for India. Table 1 gives details of stations in India whose measured data is used in the present analysis. ARN measurements in India were made at New Delhi, Gauhati, Vizag and Trivendrum during the years 1957-65 and the noise values were estimated for Pune and Nagpur using thunder storm activity data and some limited amount of measured data at these two locations *(CCIR, 1966; Ghosh and Mitra, 1958; Ghosh et al., 1974).*

Table 1
Details of Stations Whose Data is Used in the Present Analysis

Station	*Latitude*	*Longitude*	*Period*
New Delhi	28. 18° N	77. 13° E	1957-61, 1987-88
Gauhati	26. 10° N	91. 40° E	1961-65
Nagpur	21. 05° N	79. 0° E	–
Poona	18. 19° N	73. 33° E	–
Vizag	17. 41° N	83. 18° E	1964-65
Trivendrum	08. 9° N	76. 57° E	1961-65
Tirupati	13. 5° N	79. 5° E	!997-98
New Delhi	28. 18° N	77. 13° E	1997-98

In addition, National Physical Laboratory, New Delhi, has been measuring average levels of ARN since September 1987 at several frequencies using "Anritsu" Model ML 428 B field strength meter, employing vertical rod aerial and a strip chart recorder. The instrument satisfies CISPR and MIL standards. The ARN level recordings have been made for 16 minutes around each hour. Median, upper deciles and lower deciles values of ARN field strength in dB ìv/m at 1 KHz bandwidth have been worked out from these records. In order to check the accuracy of noise measurement at NPL the field strength meter is moved to rural areas, several kilometers away from New Delhi for radio noise measurements. No difference was found in these levels of noise measurements with respect to the values taken at NPL. ARN measurements were made around Tirupati (13.5°N, 79.5°E) and New Delhi (28.18°N, 77.13°E) regions at 2.5, 5.0, 10.0 MHz during 1997 and 1998..

Noise power is expressed by a quantity 'F_a' which denotes ARN field strength in terms of noise power from a short, vertical grounded, loss free antenna in dB above k To b. Here k is Boltzmann constant, b is effective receiver bandwidth in Hz and To is reference temperature taken as 288°K. Total power P_n in dB above 1 watt available at the terminals of loss free antenna is given by

$$P_n = F_a + 10 \log b - 10 \log k T_o \quad \text{—(3.1)}$$

$$P_n = F_a + B - 204 \text{ dBW} \quad \text{—(3.2)}$$

Where B = 10 log b and 10 log k T_o = -204

Noise power Fa read from the CCIR report 322 (1985) can be converted into noise field strength in dB μv/m by the following formula.

E_n (noise field strength in dB μv/m) = F_a – 95.5 + 20 log f (MHz)+10 log b (Hz)

DATA ANALYSIS AND RESULTS

Comparison of AIR Measured, CCIR Predicted and Estimated ARN from Thunderstorm Activity for India

Radio noise measurements in India at stations mentioned in Table 1 were made at several frequencies on seasonal basis during the years 1957-1965 and 1997-98 were centered around four 4-hourly time blocks viz., 0800-1200, 1200-1600, 1600-2000 and 2000-2400 hrs IST. Since estimated ARN values from thunderstorm activity *(Kotaki and Katoh, 1983)* were available for only autumn season, generally the worst period to consider, this comparative study is restricted to data of autumn months. Figures 1 to 6 show diurnal plots of AIR measured values for stations mentioned in Table 1 at 2.5, 5 and 10 MHz together with corresponding CCIR predicted values and those estimated by Kotaki and Katoh (1983) from thunderstorm activity data. Hereafter in this chapter these estimated values by *Kotaki and Katoh* will be referred to as simply "estimated values".

Following important observations can be made from this set of figures.

(i) The AIR measured radio noise values are higher at all the stations when compared to both CCIR predicted and estimated values at all the three frequencies particularly during day time hours.

(ii) The diurnal variation that can be observed from AIR measured ARN values is much lower than that seen from CCIR predicted or "estimated values".

(iii) The AIR measured values do not exhibit the dramatic increase in noise levels especially at 2.5 and 5 MHz during transition period (1600-2000 hrs) in contrast to other predicted or estimated values shown in the figures.

(iv) The CCIR predicted and estimated values show better agreement especially during daytime hours; however, during transition periods CCIR predictions can be higher by as much as 15 dB at 2.5 MHz and 5 MHz than estimated values.

It can be also observed from the figures that ARN levels are in general higher at Gauhati, Vizag and Trivendrum when compared to New Delhi. For example AIR measured ARN levels at 2.5 MHz are around 75 dB during night time (1600-2200) for the above three stations while it is around 65 dB for New Delhi. Similarly the ARN levels during daytime are also higher by 7-10 dB at these three stations when compared to New Delhi. Vizag and Trivendrum being coastal stations the observed high ARN levels there can be attributed to high levels of thunderstorm activity that is typical to such locations. Gauhati located in the north eastern region of India is also known to experience high levels of thunderstorm activity especially during pre-monsoon and monsoon months *(Ghosh et al., 1975).* In comparison New Delhi which is in northern plains of India is known to experience thunderstorm activity to a lesser extent.

Comparison of AIR Measured and CCIR Predicted Noise Values

The differences in the AIR observed and CCIR predicted ARN values are shown in Table 2 for different stations and frequencies for two local time slots viz., 0800-1200 hrs and 2000-2400 hrs. representing day and night time periods respectively. The range of values under each column represent the variation over the time interval considered. It can be seen from the table that the differences can be as high as 39 dB during daytime for Gauhati at 2.5 MHz and 19-23 dB at 10 MHz. The differences between the measured and CCIR predicted values for night time are comparatively lower and can vary between 1 and 20 dB depending on the location and frequency. The CCIR predicted ARN values are in general lower for daytime hours particularly so at 2.5 and 5 MHz. This table clearly demonstrates the extent to which CCIR predictions need to be corrected by a radio engineer while planning the HF links operating in these regions.

Table 2
Comparison between Measured and CCIR Predicted Atmospheric Radio Noise Levels in India

	Difference between observed and CCIR predicted values (dB)					
	2.5 MHz		*5 MHz*		*10MHz*	
Station	*Day*	*Night*	*Day*	*Night*	*Day*	*Night*
New Delhi 28.38° N	31-35	1-7	24-26	3-7	11-151	4-6
Gauhati 26.10° N	38-39	9-14	28-34	14	19-23	11-12
Nagpur 21.05° N	28-36	7-8	26-28	5-10	11-131	11-13
Pune 18.19° N	30-31	3-4	24-27	3-4	9-12	2-8
Vizag 17.41° N	24-31	4-8	20-30	11-12	10-13	9-12
Trivendrum 8.29° N	22-30	9-21	21-31	16-19	14-16	11-20'

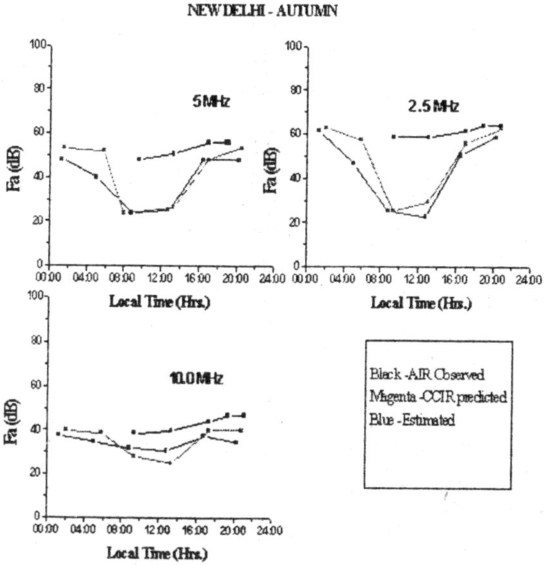

Figure 1: A comparison of observed values of ARN levels [(Fa(dB)] by AIR with CCIR predicted and "estimated values" from Thunderstorm Activity Data (Kotaki & Katoh, 1983) for New Delhi (28°38′ N ; 77° 13′ E) for 2.5, 5 and 10 MHz for Autumn months

Ghosh and Saksena (1978) have derived contour maps representing upper decile values of average ARN levels at 5 MHz in dB m v/m at 1 kHz bandwidth using measured AIR data mentioned in Table 1. They have also suggested formula for converting these noise values at 5 MHz to values at any other frequency in the HF range. In the present study an attempt has been made to compare the ARN levels for 2.5 and 10 MHz computed using the above formula and contour maps for 5 MHz with CCIR predictions. These comparisons for different frequencies are shown in Figure 7. for New Delhi for autumn. It can be appreciated from the figure that the

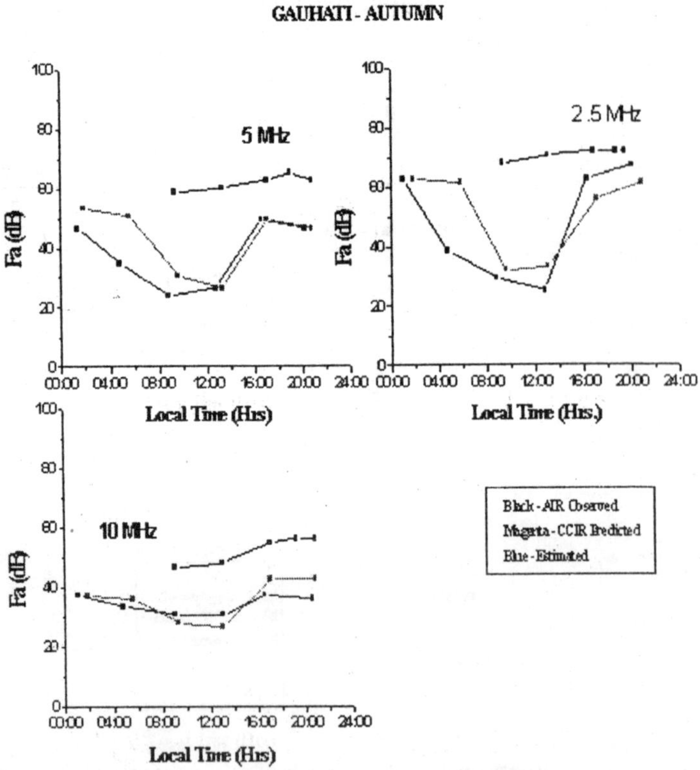

Figure 2: A Comparison of Observed Values of ARN Levels [Fa(dB)] by AIR with CCIR Predicted and "Estimated Values" from Thunderstorm Activity Data (Kotaki & Katoh, 1983) for Gauhati (26°10′ N ; 91°40′ E) for 2.5, 5 and 10 MHz for Autumn months

values predicted from the contours are much higher than CCIR predictions for all the frequencies reflecting the results shown in figure 1 and Table 2. It may be seen that the day-time differences between the values predicted from the contours and *CCIR-322* range from 24-27 dB at 2.5MHz, 17-22 dB at 5 MHz and 6-18 dB at 10 MHz. During the transition period (1600 – 2000 hrs.) the differences range from 4-6 dB on all the frequencies. Daytime differences are very large and decrease with frequency.

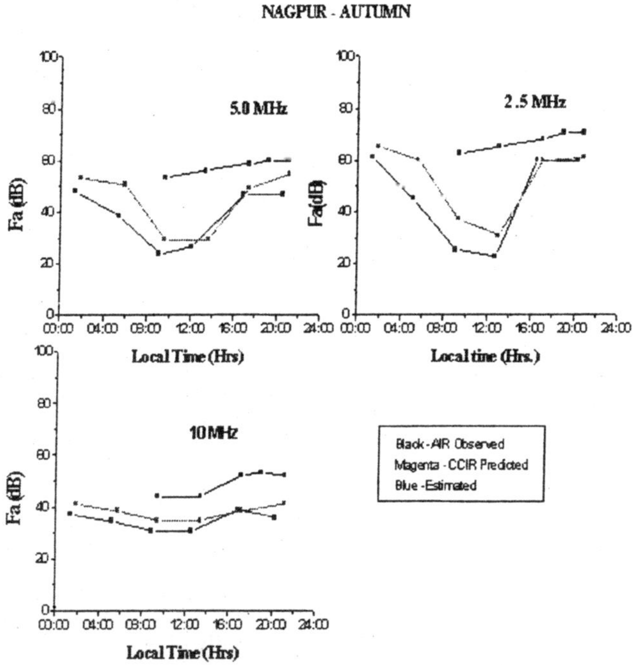

Figure 3: A Comparison of Observed Values of ARN Levels [(Fa(dB)] by AIR with CCIR Predicted and "Estimated Values" from Thunderstorm Activity Data (Kotaki & Katoh, 1983) for Nagpur (21°05 N ; 79° 0′E) for 2.5, 5 and 10 MHz for Autumn Months

Comparison of CCIR Predicted ARN Level and the Values observed by Aiya's Method

Aiya (1954) had suggested that measurement of quasi peak value of radio noise is more appropriate in tropical regions in

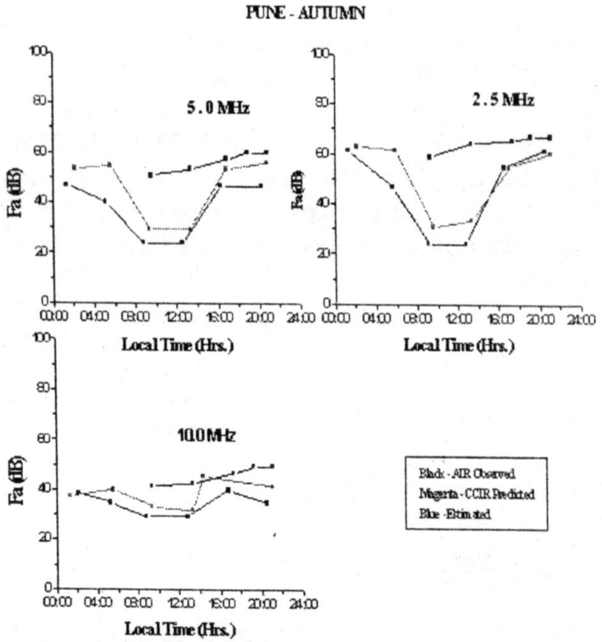

Figure 4: A Comparison of Observed Values of ARN Levels [Fa(dB)] by AIR with CCIR Predicted and"Estimated Values" from Thunderstorm Activity Data (Kotaki & Katoh, 1983) for Pune (18° 19′ N; 73° 33′ E) for 2.5, 5 and 10 MHz for Autumn months

view of peaky nature of radio noise prevalent in these latitudes and accordingly *Agrawal (1983a)* has taken noise measurements at 2.5 and 5 MHz at 6 kHz band width at New Delhi using Aiya's method during autumn months of 1970. *Satyam (1962)* has measured noise at 3 and 4.6 MHz at 6 kHz bandwidth at Bangalore during 1960-61 by Aiya's method and normalized them to 2.5 and 5 MHz. All the above mentioned measured values have been further normalized for 1 kHz bandwidth in this study and compared with the corresponding CCIR predicted values. Figures 8 and 9 give these comparisons for Bangalore and New Delhi respectively. In case of New Delhi AIR measured values have also been plotted for comparison. It may be seen from figure 8 that the differences between the measured values and the corresponding CCIR predicted values during the day-time are as high as 42 dB. During the

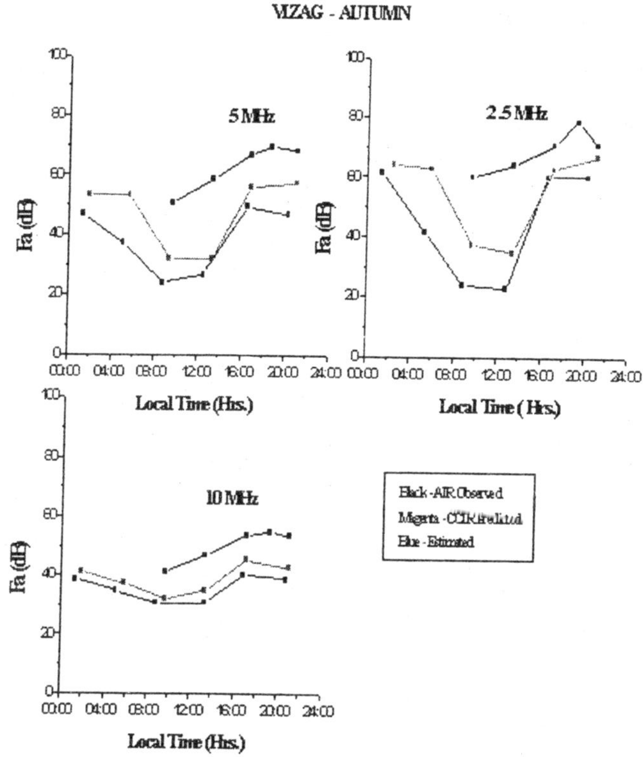

Figure 5: A Comparison of Observed Values of ARN Levels [(Fa(dB)] by AIR with CCIR Predicted and "Estimated Values" from Thunderstorm Activity Data (Kotaki & Katoh, 1983) for Vizag (17°41′ N ; 83°18′ E) for 2.5, 5 and 10 MHz for Autumn Months

transition period the differences for 2.5 and 5.0 MHz are around 10 and 12 dB from CCIR values respectively. In case of New Delhi as can be observed from figure 9. the differences are around 16 dB at 2.5 MHz and around 12 dB at 5 MHz during transition periods. These differences are due to the peaky nature of noise bursts which is accounted for in Aiya's method of measurement.

ARN Measurements at NPL

Figure 10 shows ARN values measured at NPL, New Delhi during 1987-88 at 2.5 MHz for all the four seasons along with

AIR measurements (1957-65) and CCIR predictions. Figures 11 and 12 depict similar data for 5 and 10 MHz respectively.

It is obvious from these figures that NPL measurements during 1987-88 continue to be much higher than CCIR predictions and also marginally higher (2 to 6 dB) than earlier measurements made in 1957-61 by AIR. This is true for all the three frequencies and for all the seasons. Figure 13 shows diurnal plots of NPL measurements for 3, 5, 7 and 9 MHz for summer 1988 along with AIR measurements and CCIR predicted values. This figure further confirms that the

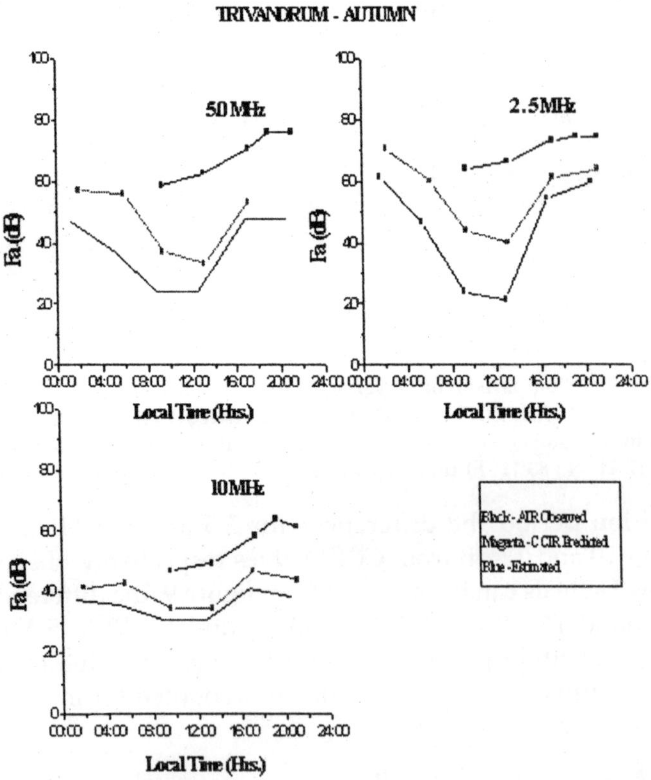

Figure 6: A Comparison of Observed Values of ARN Levels [Fa(dB)] by AIR with CCIR Predicted and"Estimated Values" from Thunderstorm Activity Data (Kotaki & Katoh, 1983) forTrivandrum (08° 29′N, 76° 57′ E) for 2.5, 5 and 10 MHz for Autumn months

observations made above from figures 10 and 11. However, the most striking feature to be noticed here is the marginal diurnal variation (5 dB) in the measured values in contrast to more than 30 dB variation seen in CCIR predictions at some frequencies.

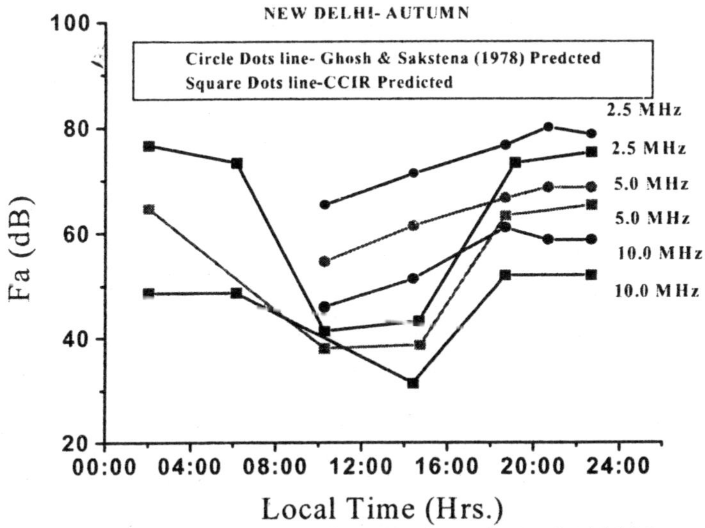

Figure 7: A Comarison of Predicted Values of ARN [Fa(dB)] from Contour Maps of Ghosh and Saxena (1978) with CCIR Predictions for 2.5, 5.0, 10.0 MHz for New Delhi During Autumn

ARN Measurements at Tirupati

As a part of this research work activity ARN measurements have been made for the first time around Tirupati region at 2.5, 5.0 and 10.0 MHz and also New Delhi region. The measurements were made in rural areas, which are free from man-made radio noise. Signals were recorded using ML 428B interference field strength meter, which measures electrical field strength and voltage signal waves and interference waves from 9 KHz to 30 MHz. With an incorporated GP-IB interface, this instrument can be externally controlled by a PC in order to process the measured data. Data has been collected during November 1997-April 1998 and were centered around four

4-hourly time blocks viz., 0800-1200, 1200-1600, 1600-2000 and 2000-2400 hours IST. The measured values have been compared with ITU-R predicted values and are given in Table-3. It is obvious from the Table 3 that the daytime values in particular are much higher than the ITU-R predictions as compared to night time values. This result is in agreement with earlier results obtained for other locations.

Table 3
Difference Between Observed and ITU-R Predicted Values (DB)

		2.5 MHz	*2.5 MHz*	*5.0 MHz*	*5.0 MHz*	*10.0 MHz*	*10.0 MHz*
Station	*Season*	*Day 12-14 hrs*	*Night 20-22 hrs*	*Day 12-14 hrs*	*Night 20-22 hrs*	*Day 12-14 hrs*	*Night 20-22 hrs*
Tirupati	Winter	23-16	08-10	20-14	05-09	08-07	05-09
	Equinox	17-16	05-06	19-19	03-01	–	–
New Delhi	Summer	22-18	09-10	23-21	13-13	19-21	15-12
	Winter	34-33	12-18	25-25	10-09	10-08	10-07
	Equinox	34-32	08-04	26-26	05-01	11-11	06-02
	Summer	33-31	06-11	26-22	09-06	18-17	04-07

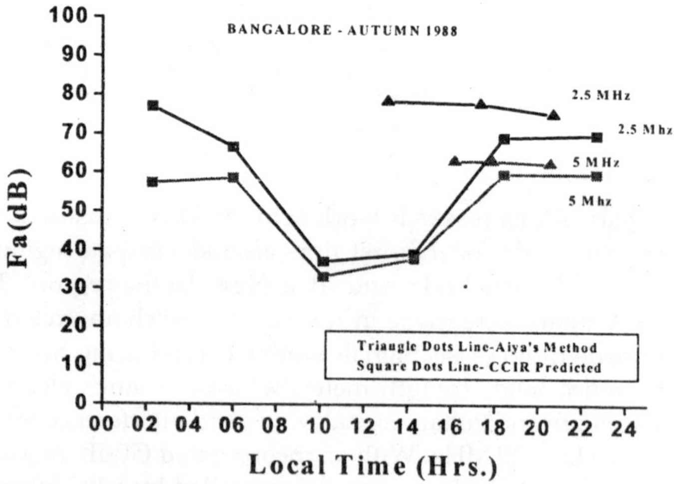

Figure 8: A Comparison of Predicted Values of ARN [Fa(dB)] from Contour Maps of Ghosh and Saxena (1978) with CCIR Predictions for 2.5, 5.0, 10.0 MHz for Bangalore During Autumn

The present study shows clearly the disparities that exist between measured ARN values and the latest predictions from ITU-R. The ITU-R map in general shows a large diurnal variation for all the locations on the globe including tropical latitudes. However, actual ARN measurements from Indian locations show only a marginal diurnal variation. This can be observed from figure 14

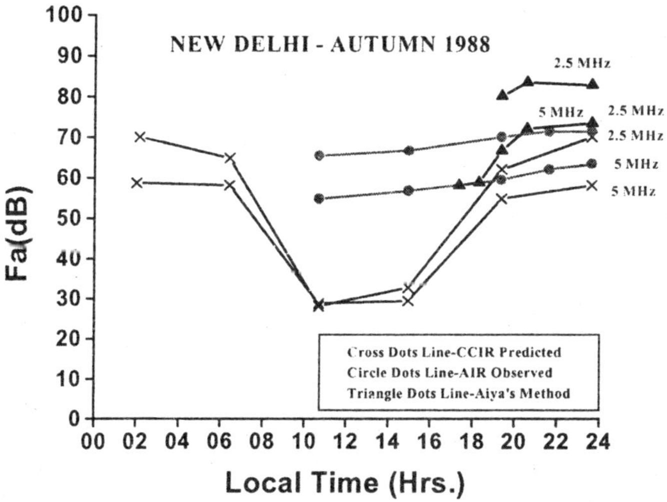

Figutr 9: A Comparison of Predicted Values of ARN [Fa(dB)] from Contour Maps of Ghosh and Saxena (1978) with CCIR Predictions for 2.5, 5.0, 10.0 MHz for New Delhi During Autumn

DISCUSSIONS & OBSERVATIONS

The present study unambiguously reveals some major disparities that exist between the actual measured values of radio noise levels over India in HF bands and those predicted by *CCIR-322* data. Prevalence of very high levels of noise during day-time hours and marginal increase in noise levels during transition periods (1600-2000 hrs) from their day-time levels are two important aspects of Radio noise over the Indian region which have not been reflected truly by *CCIR-322* noise maps. Global maps of radio noise derived from thunderstorm activity (*Kotaki and Katoh, 1983*) also do not exhibit these features of radio noise in Indian Zone. It is in general seen, the global ARN maps of both *CCIR-322* as well as those of *Kotaki and Katoh (1983)* underestimate the noise levels for the Indian region during daytime. It was also reported that estimated noise levels from thunder storm activity are in better agreement with actual measurements when compared to *CCIR-322* data especially for stations situated beyond tropics; however, in case of Singapore a tropical station CCIR predicted values have been found to be closer. A comparison of measured values in India with CCIR predicted and estimated values further confirms this observation made in regard to tropical region (Figures 1 to 6).

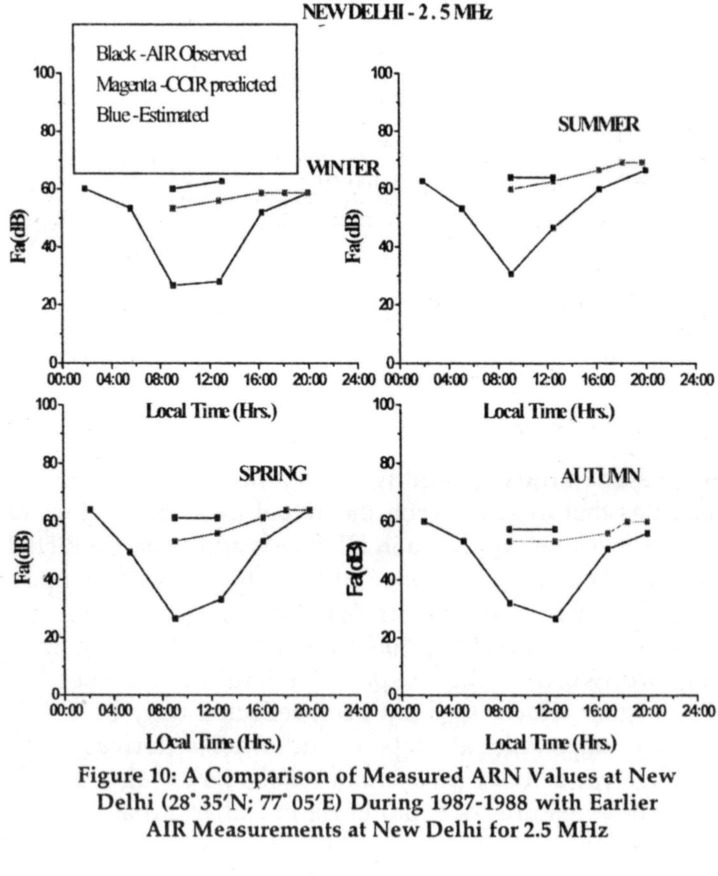

Figure 10: A Comparison of Measured ARN Values at New Delhi (28° 35′N; 77° 05′E) During 1987-1988 with Earlier AIR Measurements at New Delhi for 2.5 MHz

Kotaki and Katoh (1983) have estimated ARN levels based upon the global distribution of thunderstorm activity as observed by the ISS-b satellite and the geographical resolution of the distribution is around 1000 km x 1000 km. Accordingly the local characteristics of the lighting activity is averaged over the cell size. This area is too large and cannot give precise information about the localized thunderstorm activity confined to a particular region, as required for different regions of tropical India for correct estimation of ARN levels. It is rather difficult to estimate even statistically a thunder flash in a particular region which is the radiator of atmospheric radio noise. Accordingly there is genuine possibility of incorrect information regarding the number of flashes counted by the ISS-b. India is situated in the center of thunderstorm activity region and ARN level at any location is the function of this activity. A correct estimate of this activity can be made by the proper counts of thunderstorm flashes in a particular region, this perhaps could not be possible by ISS-b.

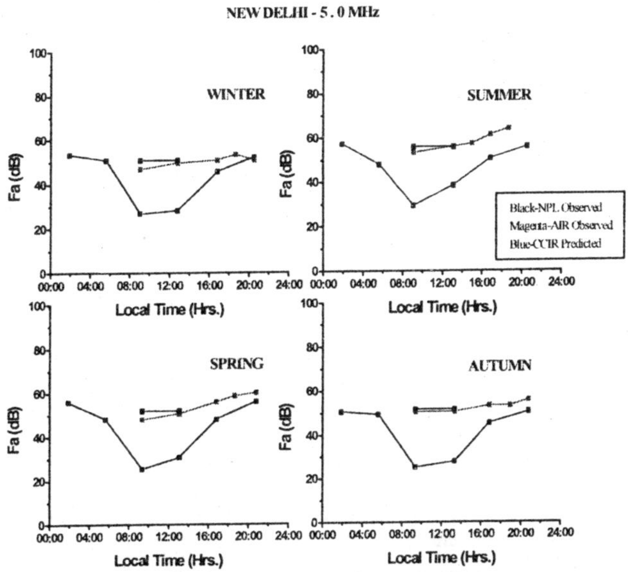

Figure 11: A Comparison of Measured ARN Values at New Delhi (28°35′N ; 77° 05′E) During 1987-1988 with Earlier AIR Measurements at New Delhi for 5 MHz

CCIR maps as well as estimated values of ARN from thunderstorm activity data show large increases of the order of 20-50 dB during night time from their day time value and these increases are generally attributed to propagated component of noise from distant thunderstorm. However, ARN measurements over the Indian region do not reveal such large increases in their nighttime values at least in 2.5-10 MHz range and they are usually of the order of 3-5 dB. The large differences (increases) between night and daytime values of CCIR data arise primarily due to remarkably low daytime values which in fact compare very poorly with actual measurements in India. Further, the observed diurnal variation of only 3 to 5 dB in the frequency range of 2.5-10

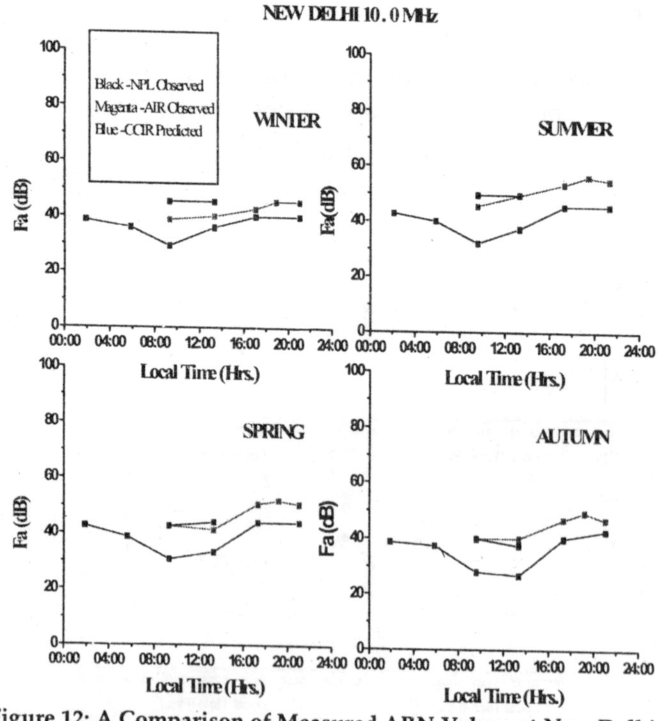

Figure 12: A Comparison of Measured ARN Values at New Delhi (28° 35′N ; 77° 05′E) During 1987-1988 with Earlier AIR Measurements at New Delhi for 10 MHz

MHz also may be because the noise component due to local thunderstorm activity is overwhelmingly large and dominates the propagated component of noise from distant thunderstorm events during night time in 2.5-10 MHz range. It was also reported by *Kotaki and Katoh (1983)* that the local component and propagated component become comparable around 20 MHz during nighttime and at lower frequencies the propagated component accounts for more than 90% of the ARN level.

The large daytime values of ARN and the modest increases from the daytime to night time values observed both by AIR and NPL especially at the lower frequencies need

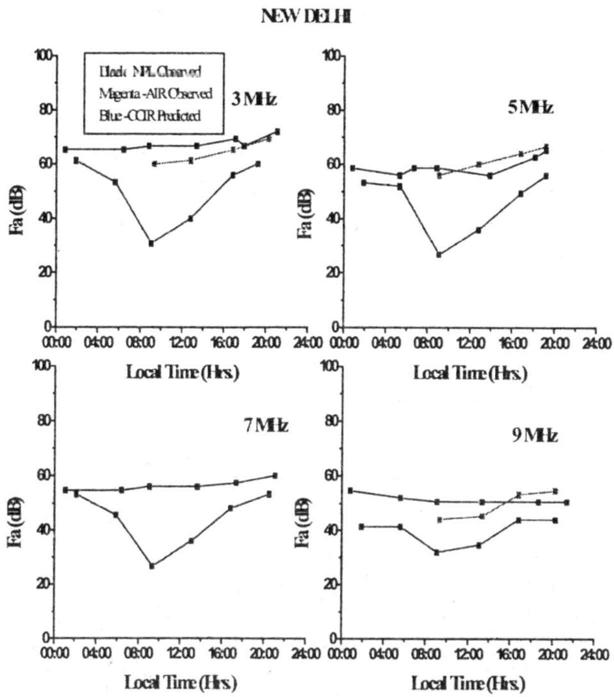

Figure 13: Diurnal Plots of ARN Measurements at New Delhi (28° 35′N ; 77° 05′E) at 3, 5, 7 and 9 MHz During Summer Months of 1988 Along with Earlier AIR Measurements and CCIR Predicted Values

a closer inspection. The possibility of contamination from man-made noise has been ruled out by conducting special campaigns at remote rural areas which also confirmed the general trend. The obvious suggestion seems to be preponderance of the line of sight component. However, there is no such obvious reason as to why it should be so at New Delhi. The possibilities are very large cloud height which will make the LOS range large enough to make the LOS component overwhelming. A second but less attractive reason is that some kind of scatter due to turbulence, layers, etc. that might extend the range of the nights propagated through troposphere. The prevalence of such large layers and glints of course is well known in tropics; but much more data, especially the frequency variation of ARN would be required before arriving at any positive conclusion.

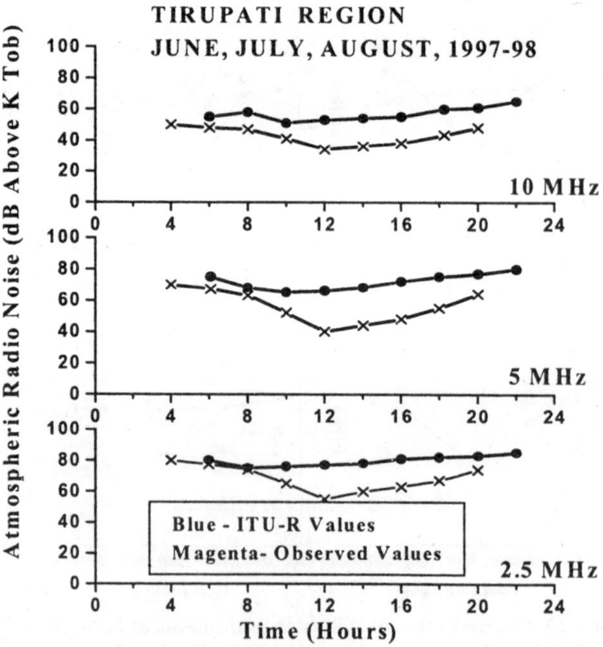

Figure 14: A Comparison of Observed Values of Atmospheric Radio Noise with ITU-R Values for the Tirupati Region

It is also necessary to study the ARN levels over India vis-a-vis thunderstorm activity to assess the contribution of local as well as distant thunderstorm activity to noise levels at locations in the tropical region. Coming to *Aiya's method (1954)* of measurements of ARN level, it may be mentioned that it measures quasi-peak value of ARN which is always found to be higher than the average observed value of noise by All India Radio or by the ARN 2 method of the CCIR. The differences between these two methods are of the order of 10 dB with the standard deviation of 4 dB. Regarding the ARN level measurements taken at NPL during the daytime, the difference between the NPL observed values and those observed by All India Radio have been found to vary between 0 to 4 dB at 5 and 10 MHz and 2 to 9 dB at 2.5 MHz.

The diurnal variation in the atmospheric radio noise (2.5-10.0 MHz) over Indian sub-continent is in total disagreement with the variations obtained at higher latitudes. This rather puzzling phenomenon has been interpreted through the results collected during this work, to be due to path attenuation differentials between the LOS mode and the ionospheric mode from the lightning source because of significantly greater cloud height in the Indian region.

CHAPTER–4

ANALYSIS OF RADIO SIGNAL FADING IN HF CIRCUITS

INTRODUCTION

HF spectrum continues to be used in a number of countries for point-to-point links and also for broadcasting and it is especially so in all developing countries in low latitudes. The advantages in this band outweigh the lower reliability as discussed in Chapter-1. Consequently, it is essential that proper predictions are given both for optimum working frequencies as well as for powers. Using higher power than that required for a particular service will load the spectrum. Accurate prediction of transmission losses and Atmospheric Radio Noise levels will go a long way in conserving the HF spectrum which is a valuable natural resource.

Reliable ionosphere prediction models have been exclusively developed for the Indian zone by NPL *(Reddy et al., 1979*) and are being successfully used for long-term planning of operational frequencies for HF systems in India. However, for estimation of field strength in HF bands in India, generally the CCIR recommended procedures *(CCIR 1970, 1978, 1986)* are being employed. CCIR had originally come out with an Interim method in 1970 *(CCIR, 1970)* for estimation of sky wave field strength and transmission losses in the range 2-30 MHz and then with an entirely new computer based procedure (second interim method), which is more complex and lengthy, in 1978 *(CCIR 1978)*. In 1986 another method was evolved by CCIR after reviewing more than 20 field strength prediction procedures from different organizations and was provided to *WARC-HFBC (CCIR, 1986)*. These methods basically differ from each other in their formulae for estimation of ionospheric absorption.

All India Radio (AIR), New Delhi has also evolved a formula for estimation of field strength (AIR method) using ionospheric absorption measurements made in New Delhi *(Rao et al., 1969)*. Field strength measurements of some regional HF broadcast transmissions of AIR at New Delhi have indicated that correlation between the measured and the calculated values of the field strength is better with AIR method than with the first interim method of *CCIR, (CCIR,1970)*. AIR method was also found to be more suitable when propagation is via F-layer *(CCIR 1974 78 a,b,c)*. Later these measured values were also compared with the values calculated using second CCIR computer based interim method *(CCIR, 1978)* and the correlation was found to be good: however, the compared data being meager no final opinion could be arrived at. AIR method was further modified in 1986 *(Sehgal et al., 1986)* to suit E-modes.

Field strength measurements of some HF broadcast transmissions of AIR in India during total solar eclipse of 16 Feb. 1980 have indicated that *CCIR (1978)* procedure can sometimes breakdown for Indian latitudes especially in estimation of ionospheric losses *(Lakshmi et al., 1983)*. AIR, New Delhi had also made a comparison of field strength measurements made in India of a number of Indian and foreign transmitters in 3-13 MHz range covering 300-3000 km paths *(CCIR, 1978-82 a)* with those obtained using a simplified manual adaptation of second CCIR interim method (1978) and concluded that the simplified version is adequate for use in the Tropical zone *(CCIR, 1978-82 b)*.

National Physical Laboratory (NPL), New Delhi had started monitoring several HF broadcast transmissions in 1987 in pursuance of CCIR study programme 30A/6 using a

Japanese Anritsu type ML 428-B field strength meter. These circuits operate between 3 and 17 MHz bands with path lengths ranging from about 300-7,000 km (see Table 1). The field strength data collected over a period of 2 years by NPL was analyzed in this study. Results of direct measurements of absorption made earlier using A_1 technique in different parts of India have also been studied vis a vis the field strength measurement taken at NPL. In addition, data collected at Tirupati (13.5° N, 79.5° E) of transmission from BBC (Masirah) and Srilanka, during 1997-98 are also used.

Table 1
Fading Allowances (dB) for Different Reliabilities at Different Latitudes in the Indian Region

	Reliabilities		
LatitudeIn Deg.	90%	99%	99.9%
10° N	12. 5	14. 0	15. 0
15°- 18° N	11. 0	13. 5	14. 0
23°-28° N	9. 0	11. 0	12. 0
ITU-R	8. 0	–	–

Following procedures for estimation of field strength and transmission losses are examined in detail in the present study.

1. AIR method *(Rao, 1969).*
2. Second CCIR Computer-based interim method for estimating sky wave field strength and transmission loss at frequencies between 2 and 30 MHz *(CCIR, 1978).*
3. A simplified version of *CCIR (1978).*
4. WARC-HFBC method *(CCIR, 1986).*

GENERAL FORMULA FOR THE ESTIMATION OF FIELD STRENGTH

The General Formula for Field Strength E_{ts} in (dB µv/m) due to a transmitter is given by *(CCIR 1986).*

$$E_{ts} = 136.6 + P_t + G_t + 20 \log f - L_{bf} - L_i - L_c - Lm - Lg - L_h - Lz \text{ dB } (\mu v/m) \qquad \text{—(4.1)}$$

Where

- f = transmitter frequency in MHz
- P_t = transmitter power in (kW)
- G_t = transmitter antenna gain in dB; at the required azimuth and elevation angle relative to an isotropic antenna
- L_{bf} = basic free space transmission loss in dB, given by
- L_{bf} = 32.45 + 20 log f + 20 log P' where P' is virtual slant range in km,

$$P' = \left[2R \sum \frac{Sin\, d/_{2R}}{Cos(\Delta + d/_{2R})} \right]$$

where

- d = hop length of n-hop mode being equal to d = D/2n, D is great circle path in km and R is radius of earth being equal to 6371 km;
- L_i = Ionosphere absorption loss in dB
- L_c = correction factor for absorption loss for E-modes
- L_m = "above MUF" loss, for frequencies above the basic MUF
- L_m = $130 (f/f_g - 1)^2$ dB
- L_m = is independent of no. of hops but it is limited to 8.1 dB for E-mode.
- L_g = ground reflection loss at the intermediate reflections points. For one hop path $L_g = 0$ for two hop paths, $L_g = 2$ dB for three hops paths,
- L_g = 4 dB.
- L_h = auroral and other signal losses and is obtained by subtracting 9 dB from median values of Y_p as given in Tables II and III of the *CCIR report* 252-2 *(CCIR, 1978).*

L_z is taken as zero for fast and simple method of field strength prediction as that of WARC-HFBC. However, its recommended value is 1.8 dB. The value needs to be revised on the basis of comparison of the predicted and measured values of field strength of a large number of known circuits.

Various Methods of Field Strength Prediction Used in India

AIR Method: Work carried out in the Research Department of AIR (All India Radio) over a full solar cycle led to the development of "AIR-Method" of field strength estimation *(Rao, 1969).*

According to the "AIR Method", expression for the non-deviative absorption is as under:

$$L_i = \frac{635[1 + 0.0017R_{12}](Cos\chi)^{0.77}\, Sec\phi}{(f \pm f_L)^2} \qquad \text{—(4.2)}$$

where

R_{12} : twelve month running average sunspot number;

χ : solar zenith angle;

Φ : angle of incidence at the absorption layer;

f_L : longitudinal component of gyro-frequency;

$\pm$: signs refer to ordinary and extra-ordinary wave components.

As stated earlier this method was found to be suitable for tropical region for radio circuits employing single-hop mode of propagation via F-Layer. However, some tests conducted on short distance circuits using E-modes of propagation showed that "AIR Method" gave much higher values of absorption as compared with the observed data and as such it gave large discrepancies in the values of field strength. In order to correct the formula, a study was carried out by AIR and the absorption expression used in the "AIR Method" was suitably modified for E-modes of propagation *(Sehgal et al., 1982).*

On the basis of various measurements of absorption conducted in India and Sri Lanka *(Ganalingam, 1969)* a figure of 270 is adopted in place of 635 in the original expression of absorption as given above, to suit the E-mode of propagation.

Second CCIR computer based interim method (for estimating sky wave field strength and transmission loss at frequencies between 2-30 MHz.)

The first CCIR interim method *(CCIR, 1970)* is not considered in this thesis as it was superseded by the second interim method. Second method was finalized by the *CCIR in 1978*. According to this method, the combined expression for both non-deviative and deviative absorption is given as under:

$$L_a = \frac{\phi_n(f_v / f_{OE}) A_T(0,0) F(\chi)(1+0.0067 R_{12}) Sec i_{100}}{(f_{ob} + f_i)^2} \qquad \text{—(4.3)}$$

where

$f_v = f_{ob}$ Cos $i_{100,}$ where f_{ob} is oblique incidence frequency and i_{100} is the zenith angle of unrefracted oblique incidence ray projected to a height of 100 km and is equal to arc Sin (0.985 Cos Δ), Δ being the angle of elevation of the ray at the ground.

$\phi_n(f_v/f_{0E})$ = is the function of (f_v/f_{oE}) being approximately independent of location, season or solar epoch and is given by *George's (1971)* results as indicated in the Nomograph given in the CCIR Report 252-2, as also confirmed by *Samuel and Bradley (1975)*

$A_t(0,0)$: absorption term which is related to absorption A_T $(R_{12}, \chi) = A_t (0,0)$.

F (χ), where F (χ) = Cos^P (0.881 χ) where P is function of month and of modified dip latitude 1 X 1 = arc tan (1/Cos χ)

I is magnetic dip in radians and χ is geographic latitude,

f_i= Electron-gyro frequency about the vertical component of the earth's magnetic field and is taken as being positive.

In the above formula f_v, the equivalent vertical frequency has been taken from the relation.

$$f_{ob} = f_v \text{ K Sec } \phi \qquad \text{—(4.4)}$$

f being the semi-vertex angle, K is the ionospheric and the earth's curvature factor which can only be evaluated iteratively. The first estimate of f_v i.e., f_{v1} is obtained from equation 4 taking k = 1 and secf at peak height of reflection. Using this estimated value for f_v, the true and virtual heights of reflection are obtained as given below,

$$h = h_m F_2 - Y_m F_2 \sqrt{1+x^2} \qquad \text{—(4.5)}$$

$$h^I = h_m F_2 + Ym F_2 (X_F \tan^{-1} h X_F - 1) + 2 Y_E (X_E \text{Cot}^{-1} h X_E - 1) \qquad \text{—(4.6)}$$

where

$$X_F = \frac{f_V}{f_o F_2} \text{ and } X_E = \frac{f_V}{f_o E}$$

$$h_m F_2 = \frac{1490}{[M(3000)F_2 + \Delta M]} - 176 \qquad \text{—(4.7)}$$

where

$$\Delta M = \frac{0.18}{X - 1.4} + \frac{0.096(R-25)}{150}$$

where.

$$X = \frac{f_o F_2}{f_o E} \text{ or 1.7 which ever is larger}$$

$$Y_m F_2 = h_m F_2 - (h^1 F - \Delta h^1) \qquad \text{—(4.8)}$$

Where $h^1 F$ is the minimum virtual height of F layer and

$$\Delta h^1 = \left(\frac{[0.613]}{X - 1.33}\right)^{0.86} - (h_m F_2 - 104) \qquad \text{—(4.9)}$$

Here $h_m F_2$ and $Y_m F_2$ are the maximum height of reflection and semi thickness of the F_2 layer. Y_E is the Semi thickness of E layer.

The first estimate of K is evaluated by

$$K = \frac{1}{\sqrt{1 - \frac{2(h^1 - h)}{R + h_0} Tan^2 h\phi}}$$

and this value of K is used to get an estimated value f^{1}_{ob} for f_{ob} from equation 4.4 and the difference $f_{ob} - f^{1}_{ob}$ is checked after each iteration. The above process is repeated several times (each time starting with the new estimated value for f_v from equation 4.4 until the difference ($f_{ob} - f^{1}_{ob}$) is reduced to the desired accuracy.

The value of K as calculated above can be substituted to find out the value of f_v/fo E and then value of f n (f_v/f_{oE}).

In order to calculate absorption and then field strength the following information would be necessary.

(i) Active Modes of Propagation: Models giving distribution of electron density at different heights are available. In order to find the active modes of propagation, a suitable model is used. These models consist of parabolic E and F layers and give a linear increase in electron concentration with height between h_mE and a point on the parabolic F_2 layer, where the plasma frequency is 1.7 foE. Using empirical equations of model for ionospheric parameters, all the physically possible paths for angles of elevation exceeding a minimum specified value via E and F layers and those for the mixed modes are evaluated.

(ii) Prediction of Ionospheric Characteristics: Ionospheric characteristics like f_0 F_2, M(3000)F_2, and eventually MUF(D)F_2 for a particular path length D are estimated using the statistical relationships between the ionospheric parameters and the solar index 12-months running average sunspot number (R)

$$f_0F_2 = A + BR + CR^2 \qquad \text{—(4.10)}$$

A, B and C are evaluated for any particular month, hour and location by the least square method using ionospheric data of 40 stations available at NPL. Similar statistical relationships are also established for M(3000)F_2

$$M\,(3000)\,F_2 = D + ER + FR^2 \qquad \text{—(4.11)}$$

Coefficients, D, E and F are evaluated for any particular month, hour and location. Equations 4.10 and 4.11 give the required values of the ionospheric parameters f_0 F_2 and M(3000)F_2 by substituting the appropriate set of values for the coefficients A,B,C,D,E and F for the required month, hour and location and the sunspot number R.

Simplified Version of Second CCIR Interim Method for the HF Band (CCIR, 1978): The *CCIR (1978)* method of *Report 252-2 (Supplement)* has been simplified to suit the Indian region. The method takes into account transmitting antenna gain, spatial attenuation, focusing gain and deviative and non-deviative absorption 'L_a'. Sporadic E layer and above MUF losses are not considered. The value of L_a is given by equation 4.3. Also in order to make the computation of field strength simpler, the value of K in the equation 4.4 has been taken as unity. This simplified method has been designated as CCIR 1 method in this thesis whereas the *CCIR (1978)* method of *Report 252-2 (Supplement)* is called as the *CCIR 11* method. Here value of K has been calculated iteratively as described in the earlier section.

WARC-HFBC Method of Estimation of Sky-wave Field Strength in the HF Band: This method was developed by the CCIR Interim Working Party 6/12 for the Administrative Radio Conference in 1986, for the Planning of HF Bands allocated to the Broadcasting Service (WARC HFBC) after reviewing some field strength prediction methods in use by different organizations. This method is a combination of simplified Report 252-2 method (for distances 0 to 7000 km) and FTZ method developed in the Fedral Republic of Germany for path lengths between 7000 to 9000 km. As per the CCIR Document 6/299-E (1989), this is the third computer based method for estimation of MUF, sky-wave field strength, signal to noise ratio, LUF and basic circuit reliability.

This method does not consider the details of propagation mechanism and is simple and quick for computation. According to this method the value of median field strength for the mode of propagation is as under:

$$L_a = \frac{677.2(Sec i_{100})}{(f+f_H)^{1.98}+10.2}\sum_{j=1}^{n} I_j \qquad \text{—(4.12)}$$

where

- n : number of hops
- χ : solar zenith angle;
- I_j : $(1 + 0.0037\ R_{12})\ (\text{Cos}\ 0.881\chi)^{1/.3}$
- R_{12} : smoothed/running average sunspot number;
- i_{100} : arc Sin (0.985 Cos Δ), where Δ is angle of elevation,
- f : radio frequency (MHz) and
- f_H : gyro frequency (MHz).

The value of absorption 'L_a' can be calculated by using the equation 4.12.

CCIR formula for calculating absorption at night: After sunset, absorption falls to a small non zero value. Extensive studies have been carried out in Japan *(Wakai 1961, 1971* and *Wakai et al., 1971)* over a range of distances upto 14,000 km. The results *(Wakai, 1975)* show that the residual night-time absorption L^1 (f_{ob}) of the ordinary wave is given by

$$L^1(\text{fob}) = \frac{(7+0.019D)(1+0.015R_{12})}{fob^2+10}\, dB(\mu V/m) \quad \text{—(4.13)}$$

where f_{ob} is in MHz and D is in km

Need for the Study on Fading: It is well known that the instantaneous signal strength of high frequency (HF) radio wave reflected from the ionosphere, fluctuates widely about the mean value and this large amplitude fluctuation over short periods known as fading, can severely deteriorate the quality of reception. Fading in HF circuits can be virulent depending on ionospheric conditions. The generic cause of course is the interference at the receiving antenna of different wave components that had travelled different paths and is known as interference fading. One usual cause is interference between rays arriving due to single hop and multi hops between ground and ionosphere. The other is the interference between the ordinary and the extraordinary components arriving with different phases. A very common cause is due to intense irregularities (spread F) which scatter the signal diffusely rather than specularly. Travelling Ionospheric Disturbances, sporadic E layer and certain focusing effects add to the problem. Depending on the cause, fading has been classified into different types like interference fading, absorption fading etc.,

It should be mentioned briefly here that the typical and more serious types of fading caused by Spread-F related irregularities are restricted to night time, primarily between 8 P.M. and midnight (local time) and are well documented *(Arons,1977; Basu et al., 1980; Dabas et al.,).* It is now well established that these night time equatorial ionospheric scintillations are caused by post Sunset lower ionospheric instabilities of the Rayleigh-Taylor Type. The depleted density-bubbles rise into the region above the peak of the F_2 layer and may travel along the field lines to latitudes beyond the Appleton anomaly peaks and affect trans-ionospheric propagation even at C-Band.

From the nature of short-term fading observed at Tirupati (13.5^0 N, 79.5^0 E) in Southern India, lasting from a second to a few seconds, it appears to be predominantly due to multi-path interference. Earlier studies made in India *[Lakshmi et al., 1994]* also support this view. Two important characteristics of short term fading are severity and rapidity or speed of fading and samples of fading signals for a few minutes to an hour in length are considered to be adequate for studying these aspects of short-term fading. Severity of fading can be measured in terms of fading range which is defined as the difference of amplitude in dB. The value exceeds for 10% and 90% of period and fading rate is the number of positive crossings per unit time through any specific level. Another important parameter for practical application is the fading allowance defined as the amount by which the median value of calculated signal strength is raised to ensure the satisfactory reception of program for a specific period of time.

A study of mean value of signal alone is not adequate for planning of any HF communication link. *ITU-R [1990]* models and formulae in general based on global data appear to be inadequate for the Indian region primarily because of scarce data inputs from this region. This study assumes greater importance in the present day scenario because of the introduction of new digital audio modulation techniques in the HF band.

HF Communications Fading: Fading is the fluctuation in signal strength at the receiver and may be rapid or slow, in general or frequency selective. In each case it is due to interference between two waves which left the same source but arrived at the destination by different paths. Because the signal received at any instant is the vector sum of all the waves received, alternate cancellation and reinforcement will result if there is a length variation as large as a half-wave length between any two paths. It follows that such fluctuation is more likely with smaller wave lengths, i.e., at higher frequencies.

Fading can occur because of interference between the lower and the upper rays of a sky wave; between sky waves arriving by a different number of hops or different paths; or between a ground wave and a sky wave especially at the lower end of the HF band. It may also occur if a single sky wave is being received, because of fluctuations of height or density in the layer reflecting the wave. One of the more successful means of combating fading is the use of space or frequency diversity. Because fading is frequency selective, it is quite possible for adjacent portions of a signal to fade independently, although their frequency separation is only a few dozen Hertz. This is most likely to occur at the highest frequencies for which sky waves are used. It can play havoc with the reception of AM signals, which are seriously distorted by such frequency-selective fading. On the other hand, SSB (Single Side Band) signal suffer less from this fading and may remain quite intelligible under these conditions. This is because the relative amplitude of any portion of the received signal is changing constantly. The effect of fading on radio telegraphy is to introduce errors, and diversity is used here where ever possible.

DIFFERENT TYPES OF FADING

Multipath Fading

Multipath fading results from dispersion of the signal by the transmitting antenna. A number of modes propagate which have variations in phase and amplitude. These waves may interfere with each other if they reach the receiver (figure 1). The signals may travel by a number of paths which, if they arrive at the receiver and are of similar amplitude, may interfere and cause fading.

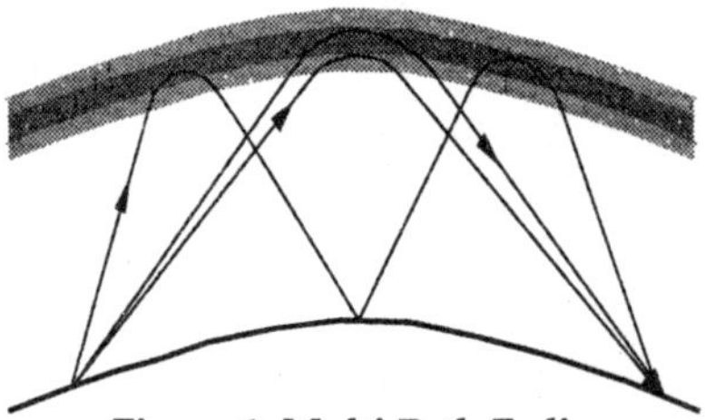

Figure 1: Multi Path Fading

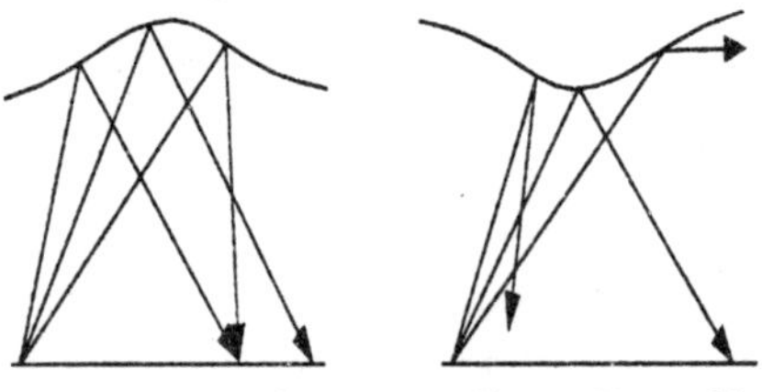

Figure 2: Focusing and Defocusing Effects Caused by Tilting and Travelling Ionospheric Disturbances (TIDs).

Disturbances known as Travelling Ionospheric Disturbances, TIDs, may cause a region to be tilted, resulting in the signal being focused or defocused as shown in figure 2. Fading periods of the order of 10 minutes or more can be associated with these structures. TIDs travel horizontally at 5 to 10 km/minute with a well defined direction of travel affecting higher frequencies first. Some originate in auroral zones following an event on the Sun and these may travel large distances. Others originate in weather disturbances. TIDs may cause variations in phase, amplitude, polarisation and angle of arrival of a wave.

Polarisation Fading

Polarisation fading results from changes to the polarisation of the wave along the propagation path. The receiving antenna is unable to receive parts of the signal; this type of fading can last for a fraction of a second to a few seconds. Skip fading can be observed around sunrise and sunset particularly, when the operating frequency is close to the MUF, or when the receiving antenna is positioned close to the boundary of the skip zone. At these times of the day, the ionosphere is unstable and the frequency may oscillate above and below the MUF causing the signal to fade in and out. If the receiver site is close to the skip zone boundary, as the ionosphere fluctuates, the skip zone boundary also fluctuates.

Short Wave Fade-outs (SWFs)

SWFs are also called daylight fade-outs or sudden ionospheric disturbances (SIDs). X-radiation from the Sun during large solar flares causes increased ionisation in the D region which results in higher absorption of HF radio waves, (figure 3). If the flare is large enough, the whole of the HF spectrum can be rendered unusable for a period of time. Fade-outs are more likely to occur around solar maximum than at solar minimum.

The Main Features of SWFs are:

(i) Only circuits in daylight sectors will be affected.

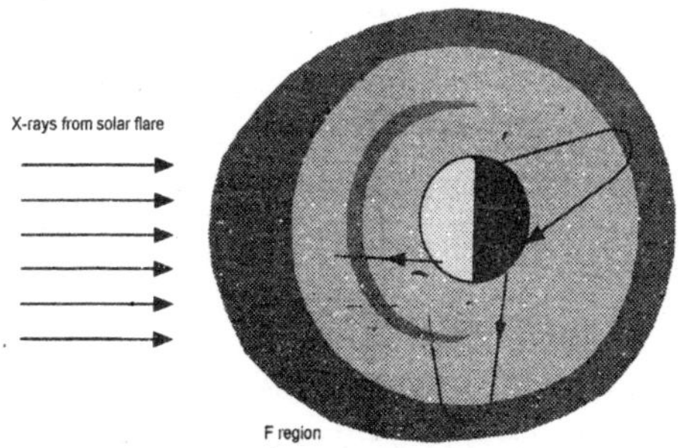

Figure 3: Fade-outs Affect only those Circuits where the Wave Passes through the D Region, that is Circuits with Day time Sectors. Night Circuits are Unaffected by Fade-outs.

(ii) Fade-outs usually last from a few minutes to sometimes two hours, with a fast onset and a slower recovery. The duration of the fade-out will depend on the duration of the flare.

(iii) The magnitude of the fade-out will depend on the size of the flare and the position of the Sun relative to the point where the radio wave passes through the D region. The higher the Sun with respect to that point, the greater the amount of absorption.

(iv) Absorption is greatest at lower frequencies, which are the first to be affected and the last to recover. Higher frequencies are normally less affected and may still be usable as shown in Figure 4.

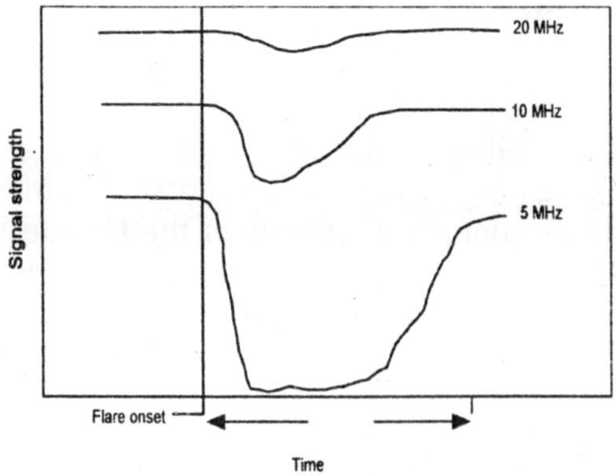

Figure 4: Fade-outs Affect Lower Frequencies First and these are the Last to Recover.

Polar Cap Absorption Events (PCAs)

PCAs are attributed to high energy protons which escape from the Sun when large flares occur and move along the Earth's magnetic field lines to the polar regions. There they ionise the D region, causing absorption of HF waves passing through the polar D region. PCAs are most likely to occur around solar maximum; however, they are not as frequent as fade-outs. PCAs may commence as soon as 10 minutes after the flare and last for up to 10 days. The effects of PCAs can sometimes be overcome by relaying messages on circuits which do not require polar refraction points; Even the winter polar zone (a region of darkness) can suffer the effects of PCAs. The particles from the Sun may actually produce a night D region.

Ionospheric Storms

Due to events on the Sun, sometimes the Earth's magnetic field becomes disturbed. The geomagnetic field and the ionosphere are linked in complex ways and a disturbance in the geomagnetic field can often cause a disturbance in the ionosphere. These ionospheric storms sometimes begin with increased electron density allowing higher frequencies to be supported, followed by a decrease in the electron density leading to the successful use of only lower than normal frequencies of the F region. An enhancement will not usually concern the HF communicator, but the depression may cause frequencies normally used for communication to be too high with the result that the wave penetrates the ionosphere.

EXPERIMENTAL SETUP & DATA COLLECTION AT TIRUPATI

Two HF broadcast transmissions, one originating from Sri Lanka (7.1° N, 79.9° E) at 11.800 MHz and the other from BBC, Masirah (20.6° N, 58.9° E) at 15.310 MHz were monitored at Tirupati (13.5° N, 79.5° E) in Southern India. For Sri Lanka–Tirupati link with path length of 720 km involving predominantly one hop mode propagation, the reflection point is around 10 N latitude close to the geomagnetic equator in the Indian zone. However, in the case of BBC–Tirupati link of 2340-km path length both one hop and two hop modes are possible with reflections occurring in the 15-18 N latitude range. These transmissions have been monitored during the period November 1997–April 1998 covering winter and equinoctial months. The data has been collected during 1000–1800 hrs, Indian Standard Time (IST) depending on the availability of transmissions. The field strength records are obtained for 5 minutes for every half an hour of the transmission period using a strip chart recorder.

Fading has been observed on a large number of occasions on both the links amounting to a total of 200 fade recordings, about 120 of them pertaining to BBC and the rest to Radio Sri Lanka. The higher number in case of BBC is because of the availability of transmissions for a comparatively longer duration (1000–1800 hrs) in a day. In case of Radio Sri Lanka transmissions at 11.8 MHz are available only during 1400–1800 hrs. These records have been analysed to derive the various fading characteristics like speed of fading, fading range and also fading allowances. Figures 5 & 6 show sample records of amplitude fading of signals at Tirupati of transmissions from BBC at 15.310 MHz and Radio Sri Lanka transmissions at 11.8 MHz. A striking difference in the two

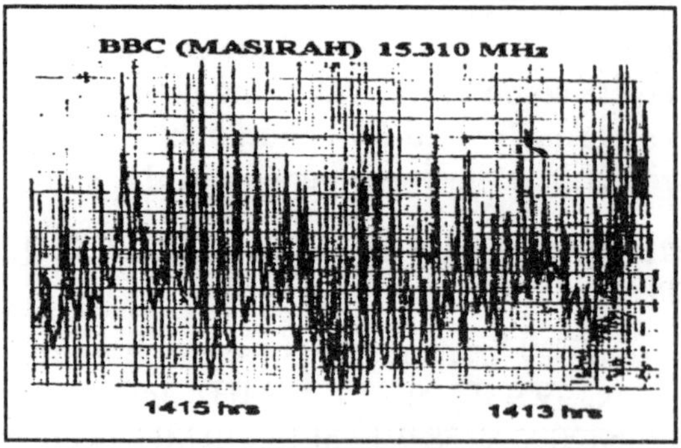

Figure 5: Sample Record of Amplitude Fading Observed at Tirupati on November 27, 1998 at 15.310 MHz (BBC).

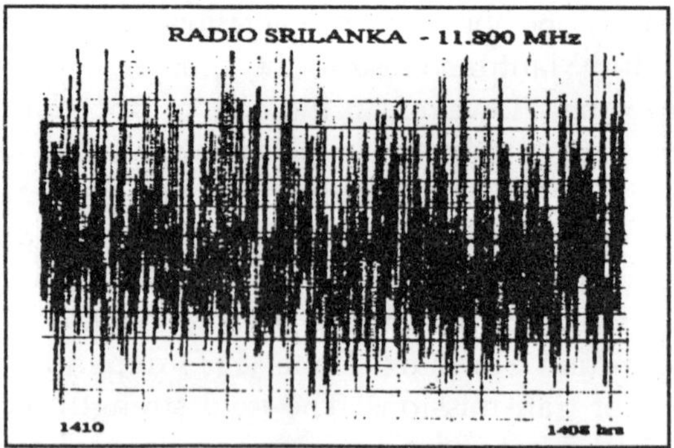

Figure 6: Sample Record of Amplitude Fading Observed at Tirupati on November 27, 1998 at 11.800 MHz (Radio Sri Lanka).

fading patterns is obvious from the above figures especially in the rapidity or speed of fading. Signals from Radio Sri Lanka seem to be fading at a faster rate as compared to those from BBC.

FADE RANGE AND ALLOWANCE

Figures 7 & 8. are histograms showing distributions of fading ranges in 2 dB intervals for Radio Sri Lanka and BBC respectively. It is observed that 40% of the 120 cases considered in this study for BBC are in 12-16 dB range and in case of Sri Lanka a similar percentage is in 14 -18 dB range indicating that fading range is comparatively greater in Sri Lanka signals. It is also observed that for 10% of the time fading range is exceeded by 22 dB incase of BBC and by 25 dB for Sri Lanka. Earlier study [*Lakshmi et al., 1994]* conducted using data pertaining mainly to 23^0-28^0 N latitude range in the Indian

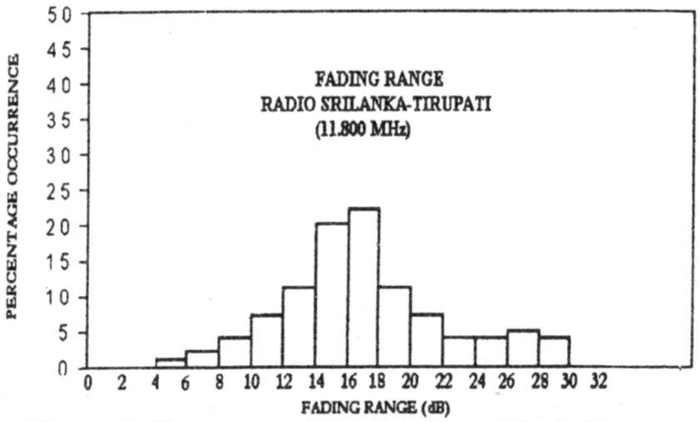

Figure 7: Percentage Distribution of Fade Range.

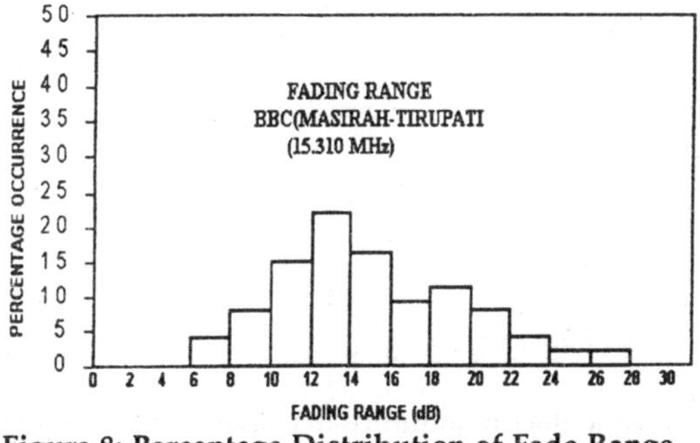

Figure 8: Percentage Distribution of Fade Range.

T able 2

Details of Various HF Circuits Monitored During September 1987 j August 1990

Transmitter Location Geog. Co-ordinates	*Receiver Location Geog- Cordinates*	*Power (KW)*	*Ant. Gain (dBI)*	*Freq. MHz.*	*Path length K m*	*Ref. Point Geog. Coordinates*	*Angle of Elevation*	*Time of Monitoring*	*MODE of Prop.*
BBC	Delhi								
(Masirah)	28.7^{0}N,	100	14	17.790	2053	24.93^0 j N,	12°	1200 hrs	One hop
20.6° N,	77.2^{0}E	-do-	-do-	15.310	-do-	67.75^0 j E	-do-	-do-	-do-
58.9°E	-do-	-do-	-do-	-do-	-do-	-do-	-do-	1500 hrs	-do-
-do-	-do-								
-do-									
Ceylon									
7.1°-N,	-do-	100	14	11.800	2407	17.85^0-N,	14^0	1300 hrs &	One hop
79.9°-E						78.63^0-E		1500 hrs	-do-
Mumbai-B5									
19.2°-N,	-do-	100	9	7.260	1136	29.92^0-N,	28.0^0	0800 hrs	One hop
72.8°-E					11.830	74.92^0-E	-do-	1500 hrs	-do-
						-do-			
Chennai-M5									
13.1°-N,	-do-	100	9	9.535	1749	20.86^0-N,	17^0	0800 hrs	One hop
80.1°-E				9.750		78.73^0-E		1500 hrs	

Contd.

Transmitter Location Geog. Co-ordinates	Receiver Location Geog-Cordinates	Power (KW)	Ant. Gain (dBI)	Freq. MHz.	Path length K m	Ref. Point Geog. Coordinates	Angle of Elevation	Time of Monitoring	MODE of Prop.
Lucknow 26.9°-N, 81.1°-E	-do-	10	7	4.880 7.250	419	27.76^{0}j N, 79.11^0-E	55^0	0800 hrs 1400 hrs	One hop
Chennai-M5 13.1°-N, 80.1°-E	-do-	100	9	9.535	902	15.75^0-N, 77.0^0-E	38^0	0800 hrs	-do-
ATA (NPL) 28.6°-N, 77.3°-E	Poona 18.5^0-N, 73.8^0-E	10	3	10.00	1200	23.92^0-N, 74.92^0-E	30^0	08 10 12 14	One hop
-do-	-do-	10	-do-	15.00	-do-	-do-	-do-	16 & 18 hrs	-do-

region has shown that these values vary between 14-19 dB depending upon the season, the higher fade ranges being for equinoctial periods.

It is normally necessary to raise the median values of the signal strength by a certain amount known as fading allowance in order to ensure reliability and satisfactory reception quality of received program for a given percentage of time (say 90% or more). Cumulative distributions of fading allowance have been obtained and shown separately for Radio Sri Lanka and BBC in Figures 9 & 10 respectively. Table 1 shows the fading allowance (dB) for different reliabilities. The values have been given separately for different latitudes (10° N, 15°–18° N) obtained from the present study and also the values from the earlier study for 23°–28° N in the Indian zone *[Lakshmi et al.,*

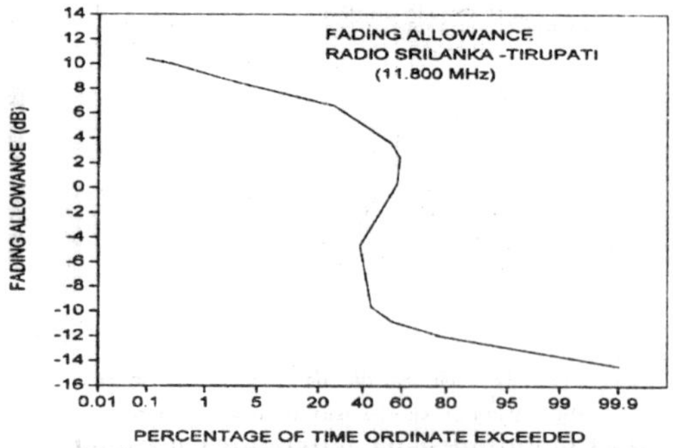

Figure 9: Cumulative Distribution of Fade Allowance-Radio Srilanka.

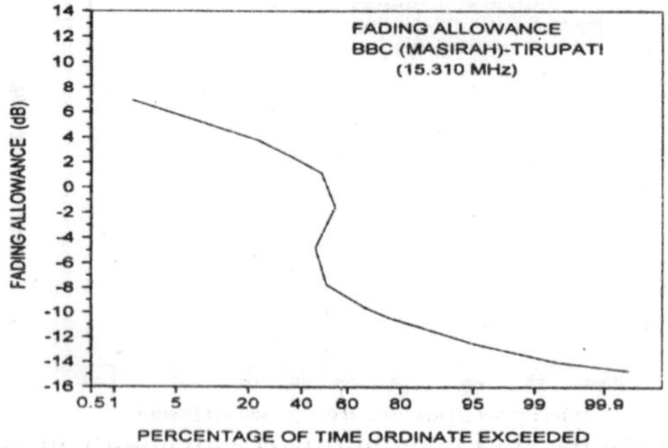

Figure 10: Cumulative Distribution of Fade Allowance-BBC.

1994]. Values recommended by *ITU-R [CCIR, 1990]* are also included for comparison. It is apparent from the Table that higher fade allowances need to be provided for HF circuits operating in 10°–18° N latitudes as compared to higher latitudes in Indian zone and have to be at least 3-4 dB higher than the ITU-R recommended values for 90% reliability.

SPEED OF FADING

Fading rates have been obtained separately for Radio Sri Lanka and BBC transmissions and the results are presented as histograms in Figures 11 & 12 respectively. The histograms show the percentage distribution in different ranges of fade rates. In case of Sri Lanka fade rates are particularly high on

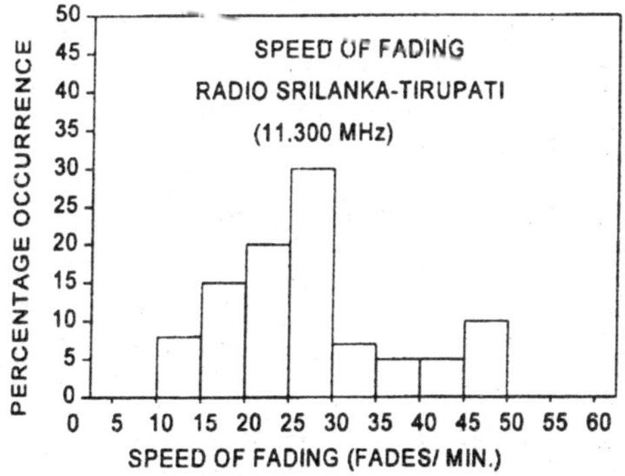

Figure 11: Percentage Occurrence of Speed of Fading.

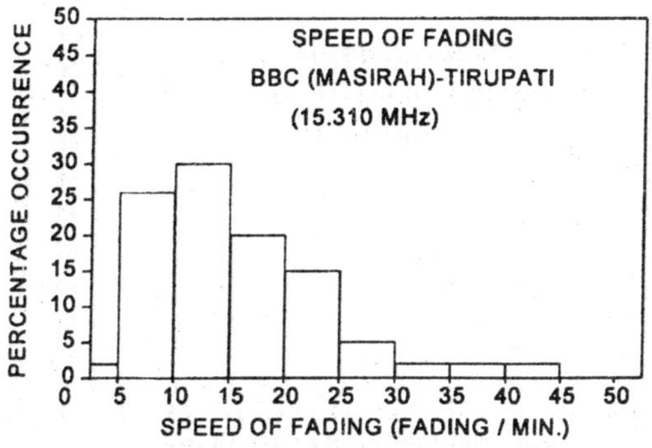

Figure 12: Percentage Occurrence of Speed of Fading.

occasions reaching values as high as 50 fades per minute and about 55% of cases have fade rates of 25 and more. A large percentage of cases (28%) occur in 25-30 fades/min range. In contrast for BBC only 8% of cases have fade rates higher than 25 and as many as 30% are in the range of 10-15 fades/min. Earlier studies *[Lakshmi et al., 1994]* using data for 23°– 28° N latitudes, have shown average rates of 14-16 fades per minute in the 9-17 MHz band and 8 fades per minute at lower frequencies of 6-7 MHz. In general speed of fading rarely exceeded 26 fades per minute.

DATA COLLECTION AT NATIONAL PHYSICAL LABORATORY, NEW DELHI

Data Base

(i) Field Strength Measurements: The basic data used for this study comprises of measured values of field strength of several HF broadcast transmissions obtained during the period September, 1987 to December 1989 at NPL, New Delhi using a Japanese field strength meter 'Anritsu' model ML428 B. The measurements have been carried out in the average position of the meter, and the detection charge and discharge time constants of the meter are within 20 ms. In addition, measured values of field strength obtained at Pune during October 1989 have also been included. In the present analysis use has been made of monthly median values of field strength. The path lengths for the various circuits chosen in this study vary from 280-2100 km. Table 2 shows the various other details pertaining to these circuits.

(ii) Ionospheric Prediction Models of NPL, New Delhi: Information on critical frequency ($f_0 F_2$) and M(3000)F_2 which contains information on F-layer peak height is necessary for computing ionospheric absorption losses. NPL, New Delhi has developed ionospheric models for prediction of HF communication parameters viz. $f_0 F_2$ and M(3000)F_2 based on a large volume of ionosonde data collected over a period of 4 solar cycles from more than 40 stations in the longitude zone 50°-150° E. These models are being used successfully by NPL

for long term planning of HF links. Use of these models have been made for predicting average monthly median conditions at control points of the various HF links considered in the present study.

DATA ANALYSIS AND RESULTS

Total Additional Losses

A radio wave undergoes a variety of transmission losses (See Equation 1) and the most important of them is due to spreading of power flux over an increasing area during its transmission (L_{bf}). Loss due to absorption in the lower layers of the ionosphere (L_i) especially in lower HF bands comes next in importance. However, as can be seen from equation 4.1 there are losses arising due to various other factors and some of which may not be either relevant or very significant to low latitudes. An attempt is made in the present study to arrive at a combined loss term (representing total additional losses in excess of free space and ionospheric absorption losses) which can account for losses arising due to scattering, spread F irregularities, focusing and defocusing effects in the ionosphere, polarization coupling and, above MUF loss L_m, correction factor for E-mode absorption losses (L_c) etc., Loss due to ground reflection in case of links involving 2 or more hops is accounted for as indicated in equation 4.1. A single term to represent the total additional losses is expected to simplify the procedure for estimation of sky-wave field strength and is particularly suitable for quick table top PC based computations.

In the present study an average value of the total additional loss for Indian latitudes has been obtained using measured values of field strength and computed values of free space and ionospheric absorption losses. Ionospheric absorption has been computed using the various procedures described earlier.

Total additional loss = $(104 + P_t + G_t - 20 \log P^{I} - Li)$

(measured field strength) in dB (μ v/m). —(4.14)

Total Additional Loss During Daytime

Field strength data collected at NPL, New Delhi during 0800-1600 hours IST for the period September 1987 to August 1989 has been used in this analysis. Monthly median field strength values for the various circuits have been obtained from the observed data. About 212 cases have been studied in detail for daytime hours. The total additional loss for each of these is computed as discussed earlier. Percentage distribution of total additional loss in different ranges computed using different formulae is shown in Figure 13. It can be seen from the figure that the most probable value of total additional loss in general is in 8-12 dB range with a mean value of 10 dB. It can be also seen that percentage (33%) number of cases in 8-12 dB range is marginally higher in regard to HFBC method. The cumulative distribution of total additional loss up to 12 dB also works out to be highest for this method with a value of 67%. This analysis also indicates that the ionospheric absorption estimated from four different formulae considered in the present analysis is fairly consistent, especially that from CCIR I and HFBC procedures.

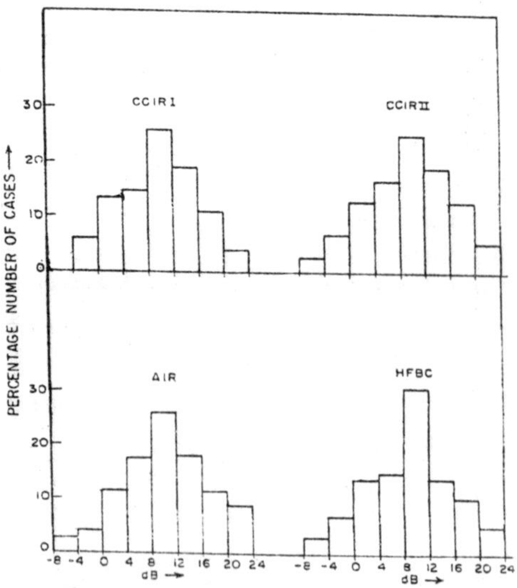

Figure 13: Total Additional Loss in dB (0800-1600 IST) for the Years 1987-1989.

Total Additional Losses During Night Time

Field strength data collected at NPL during 2130-2230 hours IST for the period September 1988 to August 1989 has been used in this study. Total additional losses has been computed using equation 13, for 38 cases and the histogram in Figure 14 shows the distribution of total additional losses for night time. It is observed that the most probable value falls in 4-8 dB range with a mean of 6 dB. The mean of additional losses in respect of 38 cases considered, works out as 4 dB.

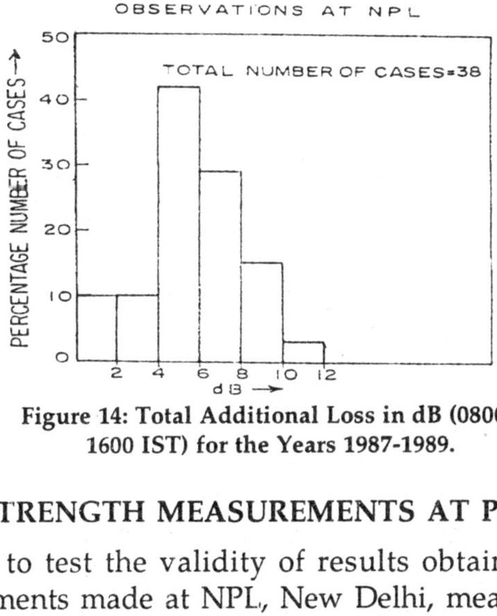

Figure 14: Total Additional Loss in dB (0800 - 1600 IST) for the Years 1987-1989.

FIELD STRENGTH MEASUREMENTS AT PUNE

In order to test the validity of results obtained from the measurements made at NPL, New Delhi, measurements of field strength of Time and Frequency Transmissions (call sign ATA) on 5, 10 and 15 MHz from NPL, New Delhi, in addition to some broadcast transmissions (See Table-3) were made at Pune during October 1989 round the clock. The total additional losses were calculated as discussed earlier, for these circuits also, however only two formulae namely HFBC and CCIRI were employed. About 7 cases were considered in this analysis. Figures 15(a) & 15(b) show the histograms of total additional loss vis a vis the percentage number of cases of monthly mean values of field strength for CCIR I and HFBC formulae respectively. The most probable value of the additional loss works out as 8-12 dB with a mean of 10 dB. These observations are in conformity with the earlier observations made in regard to radio circuits monitored at NPL, New Delhi.

DISCUSSION AND CONCLUSIONS

The severity and speed of fading are two important fading characteristics and both these aspects have been studied using two broadcast links. The two links studied here have provided information pertaining to two different locations, one around geomagnetic equator (Geomag. Lat. 0.7° N, Geog. Lat 10.2° N) and the other several degrees north of it (Geomag. Lat. 4-7° N, Geog. Lat. 15-18° N) thus enabling a comparative study of the fading characteristics of these links. A striking observation is regarding rapidity or speed of fading, the link over the geomagnetic equator exhibiting much higher fade rates as compared to the other link. The fade ranges also in general are higher for Sri Lanka. Fade allowances derived from these data do indicate that margins higher than those recommended by ITU-R need to be provided at these latitudes.

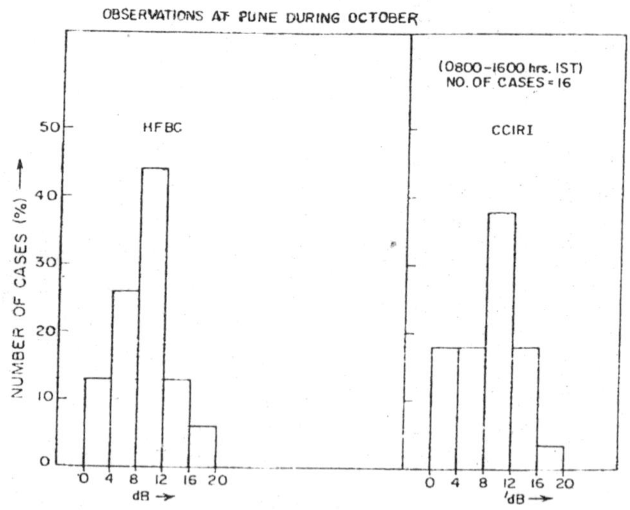

Figure 15.a: Total Additional Loss in dB(0800 -1600 IST) Oct.1989.

Figure 15.b: Total Additional Loss in dB (0800-1600 IST) Oct. 1988-89.

These results have been examined in the context of prevailing ionospheric conditions as characterized by critical frequencies and peak heights of different layers. Maximum Usable Frequencies (MUF) for different modes of propagation for both these links have been computed using available ionosonde data for this period from Kodaikanal (10° N) and Waltair (17° N) stations *[Ionospheric Data, 1997 and Rama Rao, 1998]* and also relevant information from the ionospheric prediction models developed at National Physical Laboratory for the Indian region *[Aggarwal and Shastry, 1991]* has been used.

Table 3
Ratio of Fop/MUF (D) for Different Modes of Propagation

Radio Link F_{op}	*Path Length Km*	$1F_2$	*E*	$2F_2$
BBC-Tirupati 15.310 MHz	2340	0.50 –0.56		0.70 –0.82
SriLanka-Tirupati 11.800 MHz	720	0.78 –0.90	0.90–1.0	–

In case of BBC-Tirupati link both one hop as well as two hop modes of propagation via F_2 region ($1F_2$, $2F_2$) have been considered, as they are the most dominant. However, in case of shorter path of Tirupati–Sri Lanka operating over equatorial latitudes only one hop mode of propagation via both E and F_2 layers (1E, $1F_2$) are considered. The ratios are obtained separately for each month, and the values shown in the Table 3 indicate the range of values for the entire period of observations. It can be seen from Table 3 that Fop/MUF (D) for BBC- Tirupati link is ranging between 0.50–0.56 for 1F mode and 0.70–0.82 for 2F mode of propagation resulting in favorable modes of propagation and contributing to fading. In case of Sri Lanka it is found the Fop/MUF (D) ratio is varying between 0.8–0.9 for $1F_2$ mode and for 1E mode the ratio is closer to 1 and even exceeding it on occasions, suggesting that propagation in this mode is prone for skipping and this can contribute to faster fading. It should also be mentioned here that sporadic E layer has a strong presence around 10^0 N latitude in the Indian zone with occurrence probability of more than 50% of Fop E_s values of 5.0 MHz *(Saksena, 1975)* which can contribute to strong signals when present.

The present study on fading in HF signals limited to low latitudes 10°–18° N) covering the southern zone of India indicates that fading allowances of 11-25 dB would be required for 90% reliability of HF radio signals. In an earlier study a fading allowance of 9 dB was considered adequate for 99% reliability for comparatively higher latitudes in India. Also, a pronounced increase in the speed of fading is observed for locations close to geomagnetic equator. A comparison of measured field strength in the southern region of India and ITU-R predicted values indicates that the discrepancies can be negative or positive, but, in majority of cases, the predicted values are in general higher than the observed ones. The discrepancies as large as 8 dB have been observed. The present study indicates that fading allowance of 11-12.5 dB would be required for 90% reliability of HF radio signals in 10^0 – 18^0 N latitude ranges.

CHAPTER–5

MOBILE COMMUNICATIONS-UHF SIGNAL/PATH LOSS MEASUREMENTS

INTRODUCTION

Mobile Communications

In order to meet the increasing demands for the people on move, mobile communications play a vital role. The experiments in land mobile communications were carried out in the early 1920s in Detroit where one-way broadcasts were made to receive in mobile police cars. In early 1930s mobile transmitters were developed, and the first two-way mobile system was placed in operation by New Jersey police department. World War II was an enormous stimulus to mobile communications and the end of war saw the expansion of mobile telephone services to the commercial arena. The cellular concept began to appear in Bell system proposals during the late 1940s; this idea introduced a new model for mobile radio and it is widely used today as an effective mobile communication media.

In recent decades the personal communication network has been developed to satisfy continually increasing demands for personal, local, and mobile communications *[Lee, 1986; Huges, 1989; Walfisch and Bertoni, 1988; Steele, 1989; Green, 1990; Stewart and Schaeffer, 1992; Dersch and Zollinger, 1994; Saunders,*

1994; Wei Zhang,1997; Antonio Fischer De Tohedo et al., 1998; COST 231 Final report, 1999; KIM,S.O, et al., 1999; Kurner,T, 1999; Bertoni, 2000; Andersen, 2002; Bernard De Backer et al., 2003; Shiriji Jingure et al., 2003; Janaswamy, 2003; Peng Xu et al., 2004]. To design the system successfully, it is very important to investigate the characteristics of urban radio channels, to define optimal locations of the radio ports, and to make performance predictions for the individual subscribers.

Radio signals transmitted from a mobile radio base station are not only subject to the same significant propagation path losses that are encountered in other types of atmospheric propagation, but are also subjected to the path loss effects of terrestrial propagation *(Lee, 1998)*. The general topography and roughness of the terrain tend to dissipate propagated energy, reducing the received signal strength both at the mobile unit and at base station, and thus energy terrain losses are dominant. The mobile antenna height, which is close to the ground level, contributes to this additional propagation path loss. Mobile radio signals are also affected by various types of scattering and multi path phenomena, which can cause severe signal fading. Propagation between a mobile unit and a base station is most susceptible to the effects of multi path fading phenomena, because all communication is essentially at ground level. The mobile site will receive many reflected waves, coming from different angles and direct wave, which is relatively strong compared to reflected waves. When mobile unit is moving in different directions with different speeds along its route, it passes many types of local scatterers, including a number of other vehicles. The presence of scatterers along the path constitutes a constantly changing environment, introduces many variables that can scatter, reflect and dissipate the propagated mobile radio signal energy. Heavy ground fog and extremely cold air over warm earth can cause atmospheric refraction to vary with height. However, atmospheric refraction does not effectively bend the radio path, for base station antenna is less than 91m above the ground and the transmission path is less than 23 Km *(Lee, 1993)*. Mobile radio propagation path loss is primarily due to terrain effects and the presence of scatterers along the path with in the mobile radio environment. Variations in contour and roughness of the terrain, including scatterers, cause changes in radio propagation as the result of reflections and diffractions. The transmission path between the transmitter and receiver can vary from simple line-of-sight (LOS) to one that is severely obstructed by buildings, mountains and foliages. In mobile communications, the LOS condition is considered when only reflections are present. However if diffraction of radio waves is present because of variations in terrain contour obstructing the propagation path, the LOS condition is no longer valid. For a point-to-point link, the LOS path is always present, where as in the mobile situations, it is only likely when the mobile is on a straight road leading directly away from the base station. In LOS, fast fading component is present only for a small percentage of time: a time varying factor. In the mobile situation the factor is space variant and is virtually always present *(Doble, 1996)*, due to large number of propagation paths. In LOS, flat fading is caused by bulky refractivity effects that lead to defocusing, offset launch, or arrival angles and is present for a short period. In the mobile situation, it arises from shadowing and is present for majority of time. In both cases, the aim of system planner is to obtain the maximum signal from one end of a link to the other, the transmission mechanisms and the planning involved are different. Due to the complex nature of transmission in the mobile area, the field strength prediction model developed by broadcasters cannot be used *(Doble,1996)*, and empirical methods have been developed to help the system planner. Such methods are based on determining the path loss between transmitting and receiving antennas, adding urban loss and then adding or subtracting correction factors to account for nature of terrain, urbanization extent, antenna height and street orientation.

Propagation Mechanisms

Reflection, scattering and diffraction are the basic propagation mechanisms, which impact propagation in mobile radio communications. Reflection occurs when a propagating electromagnetic wave impinges upon an object, which has large dimensions when compared to the wave length of the propagating wave. Reflection occurs from the surface of the earth and from buildings and walls. Scattering occurs when the medium through which the wave travels consists of objects with dimensions that are small compared to the wavelength. Scattered waves are produced by rough surfaces, small objects or by other irregularities. Diffraction occurs when the radio path between the transmitter and receiver is obstructed by a surface that has sharp edges. Diffraction is caused by the propagation of the secondary wavelets into the shadowed region. The secondary waves resulting from the obstructing surface are present through out the space and even behind the obstacles, giving rise to the bending of waves around the obstacle, even when a LOS path does not exist between transmitter and receiver. As the receiver moves deeper in to the shadowed region field strength decreases, the diffraction field still exists and often has sufficient strength to produce useful signal.

A sufficient number of mobile radio field strength propagation measurements were carried out abroad. In India, *Prasad and Iqbal (1998)* made VHF/UHF mobile propagation measurements by varying antenna heights. *Rao, et al., (1999)* made mobile radio propagation path loss measurements at VHF/UHF range in and around Tirupati, Southern India. *Prasad et al., (2000&2003)* have carried out mobile communication train measurements in Northern and Western India. Because of increasing demand for cellular radio service in urban areas, there is interest in making the cells very small. Also in order to reduce co-channel interference and signal delay spread, it has been proposed to place repeaters at street-lamp level. However, most of the existing knowledge of propagation loss in urban areas is based on measurements made with higher base-station antennas. In this chapter measurements made at 800 MHz (UHF Band) with low antennas in Southern India are discussed. The objective is to find the distribution of signal power on all the streets within a chosen area. The effect of individual buildings is clearly important; the buildings on both sides of the street close to the transmitter block the signal well, while open areas in other directions allow stronger signals to reach.

Path-loss Measurements

Prediction of path loss is an important factor of system design for mobile communication systems. In view of the wide geographical area and different types of terrain in this country, a single prediction technique may not be adequate for all regions. A reliable propagation model should result in path loss predictions with small standard deviation. This study helps in the design of cellular mobile systems with optimum cell sizes and transmitter powers and also mobile communication system. To identify the suitable prediction technique, field strength measurements have to be conducted in different regions and a comparison of these with prediction techniques has to be carried out. Keeping this in view, field strength measurements were carried out at 800 MHz in an urban city of Hyderabad.

The urban region where the present measurements were taken consists of building of up to 12 storeys but with very little open space between them and with street widths of the order of 10 m. The region does not consist of any skyscrapers etc. VHF or HF band mobile radios, especially for land use have attracted attention as an effective means of communications system for Police and other Government organizations. Compared with the well-known propagation characteristics on the fixed radio system, those on the land mobile service are troubled with peculiar complications. They are:

1. The antenna height of a mobile with which communication is held is very low- usually no more than 1- 3 m above ground.
2. Between the base station and the mobile, infinitely large number of propagation paths are formed due to movement from place to place.

3. This movement causes the clearance of the propagation paths to be lost. The field strength, due to the terrain irregularities and other obstacles, suffers great attenuation and location variability.

The choice of a suitable prediction technique helps to optimize coverage area and transmitter power and eliminates interference problems among various transmitters etc. All the prediction methods are divided into empirical, semi- empirical and deterministic models; the major advantage of empirical and semi-empirical method is the low computation time for large coverage areas. But at the same time these can lead to errors due to lack of precise terrain data and propagation environments and it is here that there is a decisive advantage in deterministic models. In the present study, the following prediction methods are employed. *1.Hata 2. Egli, 3.Leel, 4.COST 231 Walfisch-Ikegami 5.Walfisch-Bertoni and 6.ITU-R.* Hata's method is basically an extension of *Okumara's (1968)* method which is some what cumbersome due to numerous correction factors.

HATA'S METHOD

Hata (1980) determined the following empirical formula for propagation loss for frequencies 150 – 1500 MHz from original *Okumara (1968)* curves. The distance from base station to mobile is between 1 –20 Km. Base station antenna height is between 30 – 200 m and the vehicle antenna height is between 1 and 10 m.

In urban area

The path loss L_p (dB) can be calculated by the equation:

$$L_p = 69.55 + 26.16 \text{ Log}_{10} f - 13.82 \text{ Log}_{10} h_b - a(h_m) + (44.9 - 6.55 \text{ Log}_{10} h_b) \text{ Log}_{10} R \qquad \text{—(5.1)}$$

Where

$$a(h_m) = (1.1 \log_{10} f - 0.7)\, hm - (1.56 \log_{10} f - 0.8)$$

for medium – small city L_p represents propagation loss.

In sub-urban area

$$L_{ps} = L_p \{\text{urban area}\} - 2\{\log_{10}(f/28)\}^2 - 5.4 \qquad \text{—(5.2)}$$

In open area

$$L_{po} = L_p\{\text{urban area}\} - 4.78(\log_{10} f)^2 + 18.33\log_{10} f - 40.94 \quad -(5.3)$$

Where

h_b = Base station effective antenna height (meters),

h_m = Mobile station effective antenna height (meters),

f = Frequency in MHz, R = Distance in km.

Subsequently COST 231 Hata-Model (extended version of Hata Model (1500 – 2000 MHz) was developed. However the same is not discussed here as it is not suitable for 800 MHz.

ITU-R Model (P. 529-2, 1995)

The equation in this model has the distance term only, risen to the power b. The equation is stated as:

$$E = 65.55 - 6.16 \log (f) + 13.82 \log h_b + a(h_m) (44.9 - 6.55 \log h_b) \log d^b \quad —(5.4)$$

Where

E = Field strength, dB v/m normalized to 1 KW e.r.p.

$a(h_m)$ = $[1.1 \log(f) - 0.7]\, h_m - [1.56\log (f) - 0.8]$

b = 1 for d e" 20 Km

b = 1+[0.14+(1.87) (10-4) f + (1.07) (10-3) h_b] [log (d/ 20)] 0.8 if d>20km

These equations are meant to be extensions of the original Hata equation. When the distance is less than 20 km, which is in the useable range of the original Hata equation, the modified equation should equal the original equation.

This model claims to be suitable for use over the ranges:

Frequency range	150 - 1500 MHz
Base station height	30 - 200 m
Mobile height	1 - 10 m
Distance range	1 - 100 km

ITU-R Model P.529-3 (1999)

The equation in this model has the whole log distance term

risen to the power b. It also includes a modification to the base height term. This equation does equate to the original Hata equation for distances less than 20km. The equation

$$E = 69.82 - 6.16 \log f + 13.82 \log h_b + a(h_m) - (44.9 - 6.55 \log h_b) \log d^b \qquad \text{—(5.5)}$$

Where

a (h_m) = [1.1log (f) – 0.7] × h_m – [1.56 log(f) – 0.8]

b = 1 for d e" 20 Km

b = 1 + (0.14 + 1.87 × 10 – 4 × f + 1.07 × 10 - 3 × hbi) × [log (d/20)] 0.8 if d> 20 km

Where

hbi = hb/(1 + 7 × 10 – 6 × hb2)1/2

This model is suitable for use over the ranges:

Frequency range	150 - 1500 MHz
Base station height	30 - 200 m
Mobile height	1 - 10 m
Distance range	1 - 100 km

Lee Method

The Lee model predicts point-to-point transmission by using two components. The first is an area-to-area path loss prediction, while the second component gives the point –to-point prediction for each distance. Equation 5.6 gives the transmission loss and the constants are for 30 m base antenna height and 3 m mobile antenna height.

$$L = 142.3 - 20 \log h + 36.8 \log D \text{ (dB)} \qquad \text{—(5.6)}$$

Where h is the base antenna height in meters and D is the range in Km and the operating frequency is around 900 MHz.

Egli's Method

Egli (1957) model is systematic interpretation of measurement covering a wide span of frequencies 90–1000 MHz. In this method, the path loss (dB) can be calculated as

$$L_p = 20 \log (f) + 40 \log_{10} (R) - 20 \log_{10} (h_b) + 76.3 - 10 \log_{10} (h_m) \text{ for } h_m \leq 10 \qquad \text{—(5.7)}$$

Where

h_b = Base station effective antenna height (meters),

h_m = Mobile station effective antenna height (meters),

f = Frequency in MHz,

R = Distance in Km.

COST 231 Walfisch-Ikegami Method

COST 231 Walfisch-Ikgami model introduced some parameters to describe the character of the urban environment, namely h_{roof} is height of buildings in meters, and w is width of roads in meters, b is building separation in meters and φ is road orientation with respect to the direct radio path in degrees. These parameters are defined in figure 1 COST 231 Walfisch–Ikegami model is based on measurements performed in the city of Stockholm.

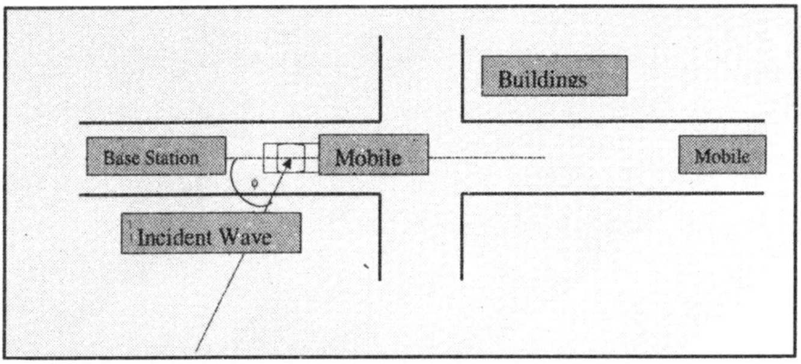

Figure 1.(b): Definition of Street Orientation.

If a free LOS exists in a street canyon, then Path loss L_b is given by

$$L_b = 42.6 + 26 \log d + 20 \log f \text{ for } d\ ^3\ 20 \text{ m} \qquad \text{—(5.8)}$$

where the first constant is determined in such a way that L_b is equal to the free space loss for d = 20 m.

Otherwise COST 231 Walfisch-Ikegami model is composed of three terms.

$$L_b = \begin{cases} L_o + L_{rts} + L_{msd} \\ L_o \text{ for } L_{rts} + L_{msd} \geq 0 \end{cases} \quad \text{—(5.9)}$$

L_o represents free space loss, L_{rts} is the rooftop–to street diffraction and scatter–loss and L_{msd} represents the multi screen loss.

The free space loss is given by

$$L_o = 32.4 + 20 \log d + 20 \log f \quad \text{—(5.10)}$$

The rooftop-to-street diffraction and scatter-loss is

$$L_{rts} = \begin{cases} -16.9 - 10 \log w + 10 \log f + 20 \log \Delta h_{mobile} + L_{ori} \\ \text{for } h_{roof} > h_{mobile} \\ 0 \text{ for } L_{rts} < 0 \end{cases} \quad \text{—(5.11)}$$

where

$$L_{ori} = \begin{cases} -10 + 0.354\, \varphi \text{ for } 0 \geq \varphi < 35^0 \\ 2.5 + 0.075\, (\varphi - 35^0) \text{ for } 35^0 \leq \varphi < 55^0 \\ 4.0 - 0.114\, (\varphi - 55^0) \text{ for } 35^0 \leq \varphi < 90^0 \end{cases}$$

$$\Delta h_{mobile} = h_{roof} - h_{mobile}, \quad \Delta h_{base} = h_{base} - h_{roof}$$

The multi screen diffraction loss is

$$L_{msd} = L_{bsh} + K_a + K_d \log d + K_f \log f - 9 \log b$$
$$0 \text{ for } L_{msd} < 0 \quad \text{—(5.12)}$$

And the terms L_{bsh}, K_a, K_d and K_f are

$$L_{bsh} = \begin{cases} -18 \log (1 + \Delta h_{base}) & \text{for } h_{base} > h_{roof} \\ 0 & \text{for } h_{base} \leq h_{roof} \end{cases}$$

$$K_a = \begin{cases} 54 & \text{for } h_{base} > h_{roof} \\ 54 - 0.8\, \Delta h_{Base} & \text{for } d \geq 0.5 \text{ km and } h_{base} \leq h_{roof} \\ 54 - 0.8\, \Delta h_{Base} .d/0.5 & \text{for } d < 0.5 \text{ km and } h_{base} \leq h_{roof} \end{cases}$$

$$K_d = \begin{cases} 18 & \text{for } h_{base} > h_{Roof} \\ 18 - 15\, \Delta h_{Base} / h_{roof} & h_{Base} \leq h_{Roof} \end{cases}$$

$$K_f = \begin{cases} -4 + ½0.7.(f / 925 - 1) & \text{for medium sized cities and suburban centers with moderate tree density.} \\ 1.5. (f / 925 - 1) & \text{for metropolitan centers} \end{cases}$$

The term K_a represents the increase of path loss for base station antennas below the rooftops of adjacent buildings. The terms K_d and K_f control the dependence of the multi screen diffraction loss versus distance and radio frequency, respectively.

Walfisch-Bertoni Method

This theoretical model was developed to predict the effect of buildings on the median transmission loss. It applies to those urban and suburban environments where the buildings are of fairly uniform height and are built in rows with small separation between neighboring buildings.

The free space path loss is

$$L_o = 32.4 + 20 \log f_c + 20 \log D \qquad \text{—(5.13)}$$

Where f_e is the frequency in MHz and D is the range in Km.

The excess path loss L_{ex} is given by

$$L_{ex} = 57.1 + \log f_c + A - \log h_f + 18 \log D - 18 \log [1 - (D^2 / 17 h_f)] \qquad \text{—(5.12)}$$

The last term accounts for the curvature of the earth, while the term

$$A = 5 \log [(d/2)^2 + (h_b - h_m)^2] - 9 \log d + 20 \log \{\tan^{-1} [2 h_b - h_m) / d]\}$$

Where h_m is the height of the mobile antenna, h_b is average building height and d is center-to-center spacing of the rows of buildings. The overall path loss L is found by adding L_{ex} to the free space loss L_o.

The simple modeling of path loss is still dominated by Hata empirical model, where the propagation results are fitted to simple analytical expression, which depends on antenna height, environment, frequency and other parameters. Hata's method is basically an extension of Okumara's method, which is somewhat cumbersome due to numerous correction factors and employs curves instead of parametric equations. These type of models are well suited for large cell area coverage systems, in which the base station is significantly elevated with respect to the surrounding environment. The Hata model does not have any of the path-specific corrections which are available in Okumara's model. The accuracy of a particular model in a given environment depends on the fit between the parameters required by the model and those available for the area concerned

(Vijay et al., 1996). In the present case we have used methods like Hata, Egli, Lee, ITU-R for comparisons. COST Walfisch-Ikegami and Walfisch- Bertoni models are only presented and are not included comparison due to lack of precise building information.

Though this type of comparison was attempted by earlier researchers *Grosskpf (1987), Aurand and Post (1988)* their conclusions cannot be adopted to our region. Grosskopf compared various prediction techniques with the measurements in hilly and mountainous terrain of Germany, the terrain was built up area with buildings of 5 floors, each floor with 4 m or so. In the present case the buildings were 2 to 12 floors in urban region. In our study we measured field strength continuously in motion covering the urban areas. Hence it is not proper to adopt his conclusions to Indian conditions. Similarly our studies are different from those conducted by Aurand and Post, their experimental studies were not mentioned explicitly. In the present study path loss was studied as a function of distance.

In all these methods effective antenna height is used. According to *ITU-R 529-1 (1994)*, base station antenna height is usually taken as the height of the center of radiation above the average terrain elevation between 3 and 15 km in the direction of interest. The antenna height of a mobile is taken as its height above ground. In the present work, there is no much variation in terrain elevation and the area is of moderate dry ground. Hence effective height of the base station $h_{eff} = h_b$ (antenna height above ground at base station) has been used for calculating path loss values. In the present study the mobile antenna height is 3 m and the average buildings height in urban areas has been taken as 10 m.

EXPERIMENTAL DETAILS

Propagation measurements were conducted at 800 MHz in Hyderabad. The signals were monitored continuously at a sampling rate of every second, averaged to one minute using a Field Strength Meter (FSM) Model R-505 of Z Technology, USA from March to May 2001. The instrument was mounted on a jeep and the field strength measurements were taken during the motion of the vehicle. Antenna of the FSM is kept at a height of 2 m above the ground. The frequency range of the field strength meter is from 3 to 1000 MHz, the measurement range is 0 dBmV to 110 dBmV and the measurement accuracy is 2 dB. It is battery operated and highly portable with LCD display. This has a facility for storage data and transfer to PC for further analysis. In the present work digital terrain data base is not used, but city topographic sheets have been used.

UHF TRANSMITTER- QUICK CALLS SMART MOBILE RADIO TRUNKING SYSTEM AT HYDERABAD

Smart Mobile is the brand name of Radio Trunking service of Motorola & Bilwara Group. Quick calls is the license holders to provide Radio Trunking Services in Chennai, Bangalore and Hyderabad. Smart Mobile is a Public Mobile Radio Trunking Service (PMRTS). Radio Trunking is a Communication tool that is ideal for group Communication with in a specified area in complete privacy. Trunking is an advanced radio communication System that uses sophisticated computerized technology to provide two way interactive mobile communication in a metro city. Smart mobile utilises SMARTNET Motorola's Radio Trunking System – which has been installed in over 23 countries and has over one million subscribers. Some of the main users in Hyderabad include HSBC, GE Capital, Delloite, DEL, Call centers and Call cabs. The specifications of the transmitter and the antenna are given in table 1.This radio trunking system transmit and receive frequencies are in the 800 MHz band. The transmit frequency of quick calls Radio Trunking System at Hyderabad is monitored for carrying out field strength/path loss measurements by moving in a vehicle in five different roads across the city.

Figure 2: Radio Trunking Transmitter Aerial Mast above Medwin Hospital Hyderabad

Table 1
Specifications of UHF Transmitter

Antenna	*Isotropic*
Antenna height	2.47 m
Tower height	18 meters
Building & tower height	67 m
Frequency	818.3375 MHz
Transmitter Output Power	60 W
Transmitter Antenna gain (G_t)	9.0 dB
Receiver antenna gain (G_r)	3 dB
Height of transmitter (h base)	87.47 m

RESULTS AND DISCUSSION

In the present study, the Quick calls Trunking System Transmitter frequency 818.3375 MHz is monitored in different environments from the transmitter. The map of Hyderabad metro city is shown in Figure 3. The transmitter is installed on a 13 storey building of Medwin Hospitals, Hyderabad, surrounded on thickly populated multistorey buildings of 5 to 15 storeys in the Chirag Ali Lane. It is surrounded on left by Lal Bahadur Stadium, Public Gardens and Assembly Hall. On the right side are Swapna lok complex with 13 storeys and other buildings of 5 to 6 storeys in the Abids road. The specifications of the transmitter are shown in Table 1. The measurements were conducted in 5 different paths which are designated as paths 1 to 5, and the typical street scenario is shown in Figures 4 (a) to (e).

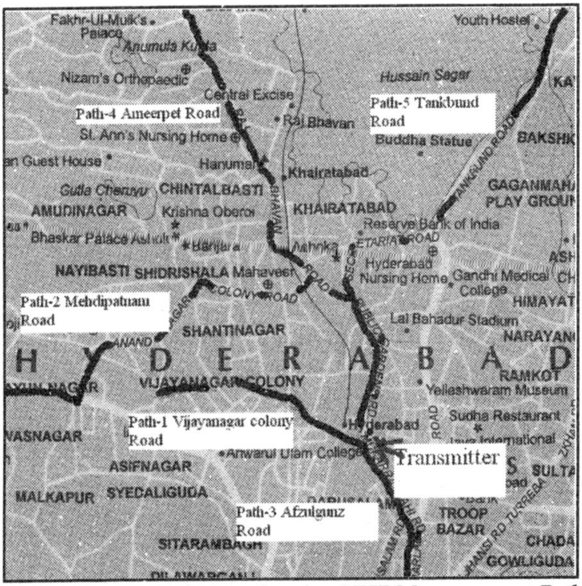

Figure 3: Hyderabad Map showing UHF Measurement Paths

Path-1-Nampalli Railway Station Circle–Vijayanagar Colony

Path 1 is a straight road of 30 meters wide, from Nampalli-Gandhi bhavan circle to Vijayanagar colony – a stretch of 5 km with buildings of 3 to 4 storyes. The first one km is densely populated with 2 to 3 story buildings and Ekminar Masjid. The zone between 1 km to 1.5 km is populated with trees and sparingly spaced buildings, Nampalli Govt. Hospital and Govt. Girls High School. The zone between 1.5 km to 2.5 km is thickly populated with residential quarters of 2 to 3 storey buildings and shops.

Figure 4.(a): Typical street scenario for UHF Measurements , Path 1. Vijayanagar Colony Road, Hyderabad

Figure 4.(b): Typical Street Scenario for UHF Measurements, Path 2. Mehdipatnam Road, Hyderabad

Figure 4.(c): Typical Street Scenario for UHF Measurements Path 3. Afzalgunz Road Hyderabad

Figure 4.(d): Typical Street Scenario for UHF Measurements Path 4. Ameerpet Road Hyderabad

Figure 4.(e): Typical Street Scenario for UHF Measurements Path 5 Tank Bund Road Hyderabad

Path-2-Nampalli Railway Station Circle–Mehdipatnam Bus Terminus

This path covers a distance of 5.5 km. with 60 meters wide road with thickly populated buildings, one flyover bridge, Army area and Bus Terminus. The zone up to 1.5 km is covered by Nampalli Railway Station, Public Garden with more trees on the left side. Hotel Harsha, Residency, Medwin Hospital, Madina Educational center, Haz House, L.B. Stadium and Police Control Room are on the right side and multistorey buildings of 5 to 7 floors. The zone from 1 km to 2 km is covered by Assembly Hall, D.G.P. Office, Lakadikaful Bus station on the left side and All India Radio, Ravindra Bharathi, Hote Harsha on the right side. All these buildings are ranging from 5 to 10 storeys height. The zone from 2 km to 5 km is covered with shopping complexes, Hotels, Offices, Colleges and parks with building heights ranging from 3 to 10 storeys. In the left side of the path in this zone–Hotel Wood bridge, A.P. State Trading Centre, A.P. Film & Theatre Development buildings, JNTU (Jawaharlal Nehru Technological University), Polytechnic College & NMDC., Premier Hospital are located. On the right side of the path in this zone – PTI Building, Mahavir Hospital, Hotel Golkonda, Nehru Children park, Sarojini Eye Hospital and Army area of 10 acres are located.

Path-3-Nampalli Railway Station Circle–Afzulgunz

This path covers a distance of 2 km. This zone consists of thickly populated buildings of office complexes, commercial complexes and Corporate Hospitals with buildings of height from 5 to 15 storeys. This path is a 60 meter wide road extending to thickly populated buildings of 3 to 5 storeys in the side lanes. Osmania General Hospital and CDR Hospitals are also located in this path.

Path-4-Nampalli Railway Station Circle–Ameerpet

This path covers a distance 6.4 km upto Ameerpet circle with 60 meter wide road. This zone is very thickly populated with Commercial complexes, Hospitals and offices. The path takes approximately "L" curved shape. Most of the buildings are 5 to 10 storeys. The zone up to 2 km. is common for path 4 and path 2. From 2 km to 3.5 km, the path takes an "L" shape and is covered with less densely populated buildings of 3 to 6 storeys. The zone from 3.5 km to 6.4 km is thickly populated with multistory buildings of 5 to 12 storeys. Some of the important buildings are Eenadu office, A.P. Transco, Medinova, NIMS and Hotel Urvasi.

Path-5-Nampalli Railway Station. Circle–Tankbund, Secunderabad

This path covers a distance of 5.5 km up to R.P. Road and S.D. Road junction circle. The zone upto 2.5 Km is covered with thickly populated offices, commercial complexes, Hotels

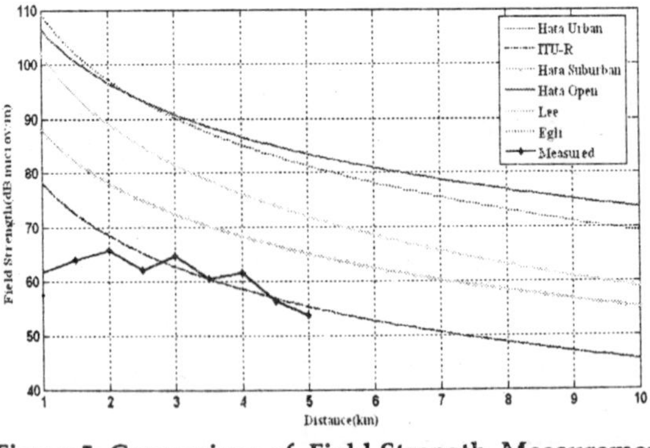

Figure 5: Comparison of Field Strength Measurements with Propagation Prediction Models (Path 1. Vijayanagar Colony Road, Hyderabad).

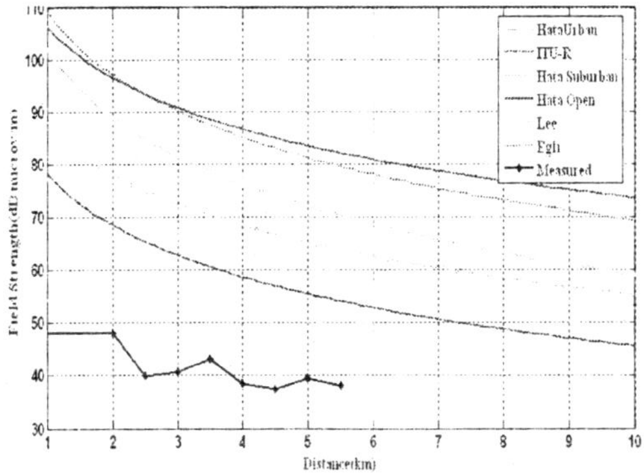

Figure 6: Comparison of Field Strength Measurements With Propagation Prediction Models (Path 2. Mehdipatnam Road, Hyderabad).

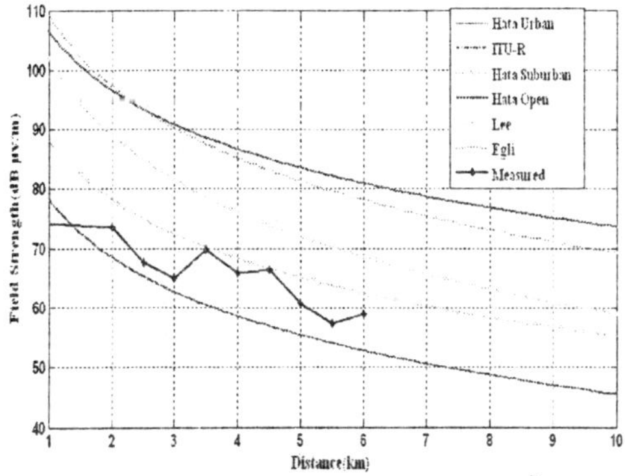

Figure 7: Comparison of Field Strength Measurements with Propagation Prediction Models (Path 3. Afzalgung Road, Hyderabad)

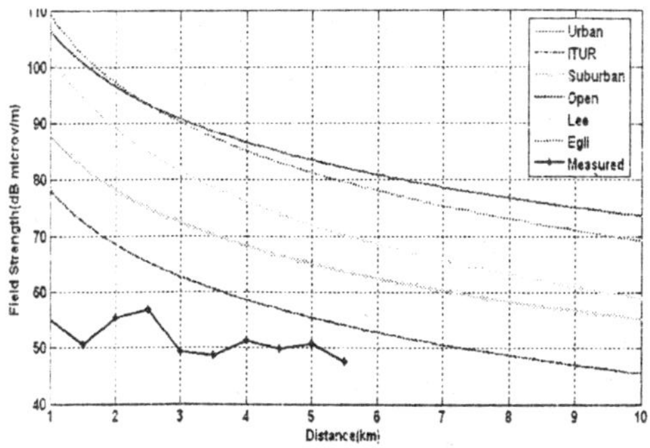

Figure 8: Comparison of Field Strength Measurements with Propagation Prediction Models (Path 4. Ameerpet Road, Hyderabad)

and Corporate Hospital Buildings of 5 to 15 storeys with a flyover leading to Hussen Sagar Tank bund. This path covers Reserve Bank of India, Secretariat, Birla Mandir, Medicity Hospital and B.R.K. Bhavan. The zone from 3 km to 5 km i.e., 2 km long road is a 60 meter wide straight road on the Hussen sagar with out any buildings. The zone from 5 km to 5.5 km is covered with sparingly placed buildings of 3 to 5 Storeys.

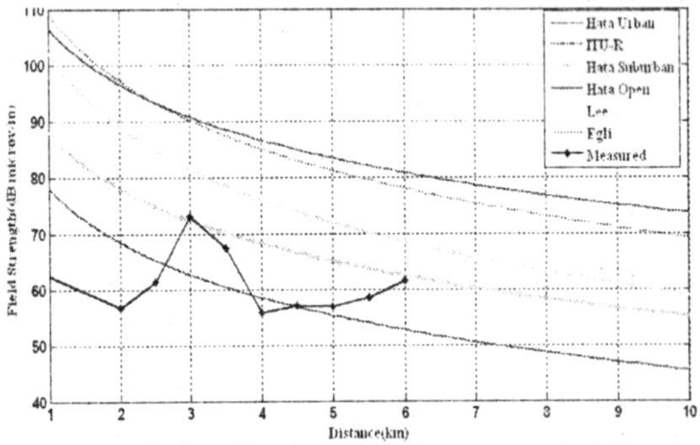

Figure 9: Comparison of Path Loss Measurements with Propagation Prediction Models (Path 5.5ankbund Road, Hyderabad)

Table 2
Standard Deviation of Errors in the Prediction Methods

Propagation Model (800) MHz	*Mean Error (dB)*	*Error of Standard Deviation (dB)*
Egli	30.55	8.082
Hata	6.982	6.02
Lee	15.00	7.134
.ITU-R	6.982	6.02

From the environmental description provided above it is very clear that most of the measurement paths fall under urban environment except the path-5, along the tank bund which falls under sub-urban. For comparison the path loss values are converted into field strength values w.r.t. to 1 KW ERP (Effective Radiated Power).Comparison with different propagation models are shown in Figures 5 to 9. Signal strengths are recorded while moving the vehicle with an average speed of 40 kmph. Further it should be mentioned here that both ITU-R model and Hata-urban models give the same predictions upto 20 kmph and the two curves merge with one another and hence ITU-R model curve is not seen separately.

In general the *Hata-Urban model (ITU-R)* gives good estimates. The standard deviation of errors in the prediction methods are given in Table 2. It can be seen from the table that in general Hata urban or ITU-R models are giving less errors in the predictions compared to the other models. This study clearly recommends the use of Hata-urban model for UHF Band planning for Radio trunking and Cellular applications. It is also pointed out here that earlier studies by *Rao et al., 1999* draw similar comparisons.

CHAPTER–6

TROPOSPHERIC SCINTILLATION STUDIES ON SATELLITE COMMUNICATION LINKS & SCINTILLATION PREDICTIONS USING RADIOSONDE DATA

INTRODUCTION

The evolution of modern satellite systems has led to a steep increase in satellite capacity, considerable reduction of onboard satellite equipment size and significant cost reduction. With the advancement of satellite technology to meet the growing demands of mobile communications across the world, the need for satellite services are continuously increasing. One of the major reasons is the rapid deployment of small size satellite communication terminals, such as Very Small Aperture terminals (VSATs) for voice, data and multimedia services (*Asoka Dissanayake, 2002).* With the use of satellite services rapidly increasing, the lower frequency bands will become congested. The congestion of primary frequency bands resulted in use of higher and higher bands from L, S or C bands to X and Ku bands, and in near future to Ka and EHF bands. The C-band (4-8 GHz) frequencies are used for satellite communications in tropical areas such as India and other Asian countries. The rapidly increasing demands for the use of geo-stationary satellites for telecommunication and

Television transmission and reception, indicate that the lower C-band frequencies would soon be used up.

For Satellite communication services in future, there is a need to deploy the higher Ku band (11-18 GHz) and even higher Ka band frequencies (20-30 GHz) for providing higher data transmission rates, reduced interference potential and requires smaller receiving stations and antennas, leading to greater mobility *[Costa, web site]*. Use of smaller earth terminals for transportable applications make it difficult to provide sufficient link margin for propagation related outages. Hence, reliable prediction of propagation impairments for low-availability systems becomes important *[Asoka Dissanayake, 2002]*. One of the main considerations while using high frequency Ku and Ka bands, is the influence of the atmosphere on radio wave propagation. Since the influence of the atmosphere on microwave propagation generally increases with the frequency, a detailed analysis has to be made, so that these higher bands can be exploited for high quality Satellite services.

Propagation impairments affecting Ku and Ka-band satellite links include, troposphere scintillation, gaseous absorption, cloud attenuation and rain attenuation. In this chapter the effect of Tropospheric scintillation on Ku Band Satellite Communication Links are studied. In particular for Southern India, long term Scintillation statistics are predicted *(Kolmogorov theory)* using Radiosonde data collected from the India Meteorological Department. This method is applied for two years of Radiosonde Data collected at Bangalore, India from 2003 to 2004 and the results are compared with scintillation measurements carried out from Aug. 2004 to November 2004 at Master Control Facility, Hassan, (13.07°N; 76.08°E) (which is 200 km away from Bangalore) by measuring signal strength of INSAT-3B Satellite (Ku Band Beacon).

TROPOSPHERE

The troposphere is dominated by a variety of circulation systems on a wide spectrum of spatial and temporal scales. The planetary and synoptic scale motions give rise to zonal symmetric circulations such as Hadley cells, monsoons, cyclonic and anti-cyclonic phenomena that are so important for our very existence. The radioactive transfer and water vapour phase transformations add fuel to the atmospheric dynamics. Troposphere extends from ground level upwards to altitudes of about 16 km at the equator and 8 km at the poles. The percentage composition does not change with altitude except that of water vapour. The troposphere is characterized by a 6°C per km lapse rate on the average. Radiation, absorption, condensation, evaporation, advection, convection turbulent diffusion and a number of complex and interacting meteorological phenomena make the troposphere a cauldron of non-linear and unpredictable processes. The thermal structure of the troposphere is important not only from considerations of atmospheric dynamics and heat exchange but also from radio propagation point of view.

Radio Refractive Index in the Troposphere

For all practical purposes, the atmosphere may be considered to consist of dry air and water vapour from refractive index considerations. Molecules of dry air do not have a significant permanent dipole moment, that is, the displacement between the effective positive and negative charges in the molecule is essentially zero. However, under the influence of an external electromagnetic field, (due to a passing radio wave) a small dipole moment is induced. The net effect of all such induced dipoles in the sphere of influence of the radio waves is to increase the dielectric permittivity of the medium and hence the refractive index. Since the density of a gas is directly proportional to its partial pressure and inversely to its temperature, the refractive index contribution of any gas is proportional to its partial pressure divided by its temperature. Water vapour in the atmosphere profoundly influences the tropospheric Radio Refractive Index. The water molecule possesses a permanent dipole moment and these permanent dipoles normally are randomly distributed. However, in response to the magnetic field associated with a passing radio wave, the permanent dipoles (water vapour molecules) partially align with such magnetic field adding to the polarisation of the medium. This effect is obviously proportional to the number density of water vapour molecules. But with increasing temperature, the random thermal motions make it difficult to discipline the dipoles to fall in line with the imposed magnetic field, thus the polarisation is inversely proportional to the temperature. With the aid of constants derived from experiments, the refractivity N (refractive index in-excess of unity multiplied by a million) can be written as

$$N = \frac{77.6}{T}\left(P + \frac{4810e}{T}\right) \qquad \text{——(6.1)}$$

where P is the dry atmospheric pressure and e the water vapour pressure, both in millibars and T is the absolute temperature.

To facilitate a clear description of the dry and wet parts of the refractivity, it can also be written as

$$N = 77.6\ P/T + \ (3.73 \times 10^5 \times e\ /\ T^2\ ) \qquad \text{——(6.2)}$$

It is obvious from the above equation that water vapour plays a dominant role in tropospheric refractivity. It should be noted however that with increasing tempera-tures, the wet part of refractivity steeply increases (contrary to what is apparent from equation 6.2) because for a given relative humidity, the water vapour pressure increases rapidly with increasing temperature. The proportion by volume of water vapour at ground level varies from near zero levels at the poles to more than 6% in the tropics. It is also interesting to note that the mass of water vapour in tropics may exceed 30 grams per cubic meter, while even during torrential rains the atmosphere has less than 10 grams of liquid water per cubic meter in extra tropical countries. The water vapour distribution, especially in tropics, assumes great significance because of its role in modifying turbulence-related refractivity fluctuations, in causing attenuation at frequencies above 15 GHz and in producing large refractivity gradients leading to anomalous propagation conditions. A typical water vapour profile for

South East Cost of India is shown in figure 1. Most of the sharp gradients usually are smoothed out in such average profiles and care should be exercised while using such data for sensitive applications.

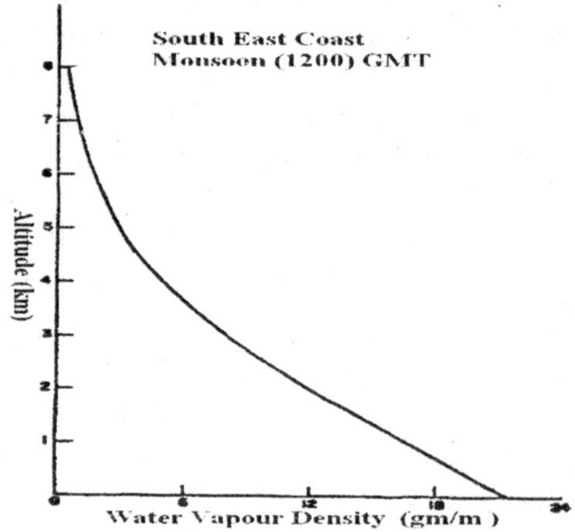

Figure 1: A Sample Water Vapour Profile at 1730 hrs Local Time from the Indian Water Vapour Atlas

By far the most cardinal parameter in predicting the course of a radio wave propagating in the troposphere is the refractivity gradient, especially in the lowest part of the troposphere. By differentiating equation (6-1) we get

$$dN/dh = 77.6\,[1\,dP/Tdh - (P/T^2+9620e/T^3)$$

$$dT/dh+4810\,de/T^2dh]\ m^{-1} \qquad \text{——(6.3)}$$

By substituting standard (average) gradients of pressure, temperature and water vapour at ground level for temperate latitudes in equation (6-3), we get a value for dN/dh as

$$dN/dh = -4 \times 10^{-2} \text{ per meter or}$$

$$-40 \text{ N per kilometer} \qquad \text{——(6.4)}$$

The curved path of the radio wave refracted in an atmosphere with such gradient is equivalent to a straight line path on an imaginary Earth with a radius 4/3 times the true radius of the Earth. Observations in India show that for an overwhelming percentage of time the initial refractivity gradient is much more negative (*Sarkar et al., 1985*).

Tropospheric Turbulence

Surface refractivity and refractivity gradient is most important for understanding tropospheric propagation, the generation of turbulence is also important in certain specific applications. In addition to propagation in LOS and diffraction modes (especially over a ridge), transhorizon propagation is possible for limited bandwidth applications using scatter of radio energy from irregularities omnipresent in the troposphere. The mesoscale phenomena caused by squall lines, mountain lee waves etc., gradually cascade (usually aided by shear forces) into smaller scales until they end up as heat due to viscous dissipation. Formation of eddies and their final dissipation is shown in figure 2. The original approach as proposed by *Kolmogorov (1941)* assumes that through non-linearity of fluid flow processes a quantity of power ° per unit mass is supplied to the turbulence at the largest turbulence scale size of the order L_0. This energy is transferred again in tandem to those eddies which dissipate the energy. If the equations were linear, such transfer between different eddy sizes would not be possible. If the lowest eddy size when

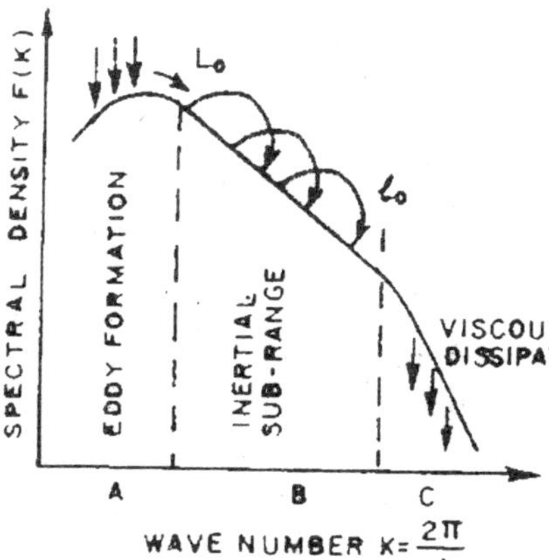

Figure 2: Generation of Atmospheric Turbulence and Cascading through Internal Sub range into Viscous Dissipation

dissipation occurs is $l_{0,}$ the energy spectrum E (K) can be derived as

$$E(K) = a_1\, \varepsilon^{2/3} K^{-5/3} \quad 2/L_o < K < 2/1_0 \qquad \text{——(6.5)}$$

where a_1 is a constant close to unity and K is the mechanical wave number (=2π/λ). At the initial stages as long as the local velocity of air is less than critical. laminar flow is obtained. At critical velocity the laminar flow turns into a turbulent flow and large eddies of size L_0 are formed as shown in figure 2. In the range demarcated B the energy is conserved but the eddies decay in size gradually and continually until the lower size limit 1_0 is reached. This region B is called the inertial region. The eddies are so small in range C that their energy is dissipated due to the viscosity of the medium. The eddies in the inertial sub-range participate in the scatter of radio waves due to the random variations in permittivity.

To summarise, turbulence is created by horizontal movement of one air stream over another, by vertical movement of moist and warm air parcels and by a process of entrainment whereby parcels of one air mass are injected bodily into a neighbouring region of different temperature and humidity. The inertial sub-range of eddies is important for troposcatter propagation as they contain blob sizes comparable to the wavelengths commonly used in scatter communication systems from a few 100 MHz upto 5 GHz. Assuming that within the inertial sub-range the refractive index fluctuations are isotropic the one dimensional space spectral function F_n (K) can be written as $F_n(K) = ¼\, C^2_n K^{-5/3}$ *(Tatarskii, 1961)*. The structure constant C^2_n is a measure of the variability of refractive index fluctuations and is an important parameter in characterizing power and antenna sizes in troposcatter links. It may be noted here that C^2_n assumes very large values in the tropics aided by high humidity and temperature fluctuations. This facilitates troposcatter systems to be very effective.

India being a tropical country, tropospheric environment is highly dynamic and is very often dominated by frequent occurrences of sub refraction, super refraction and ducting, high rain rates, multi path effects compounded by Fresnel zone clearance problems in mountainous and irregular terrains *(Narayana Rao et al., 1985)*. Some of the problems presented by the Indian troposphere to the design engineers of microwave systems of high reliability are caused by large variations of humidity at high temperature levels. At lower temperatures the variation in refractive index is only a little for whole range of humidity values from zero to saturation, in contrast to great variations in the refractive index values at higher temperatures for the similar change in humidity. In fact, higher the temperature, larger is the change in refractive index value for the same amount of humidity change *[Hall, 1979]*. Tropospheric environment over tropical region is dominated by high temperatures, humidity and rain fall rates and exhibits propagation characteristics, which are radically different from mid and high latitudes.

India has its own special climate features which are different in different parts of the country. The climate of India can be classified as follows. The winter season, particularly in the northern half of India is from October to March. April, May and June are summer months all over India. This is followed by the monsoon from July to September. In the Southern half of the Indian peninsula the winter is not severe. In the south east of the lower peninsula, the period from June to September is not very rainy as elsewhere, however from October to December this region experiences the north east monsoon associated with cyclonic storms from Bay of Bengal. There is not only a seasonal variation, but an appreciable diurnal variation. From this respect one cannot adopt the conditions prevailing in the extra tropical countries where extensive studies have been made regarding radio wave propagation.

TROPOSPHERIC SCINTILLATION CHARACTARISATION

The basic theory and the method of estimating the scintillation effects on satellite paths proposed by *Vasseur,H* (*1999*) is presented below. Tropospheric scintillation fluctuations affect satellite communications links especially those operating above 10 GHz. Under clear-sky conditions and also during rainfall, scintillations will occur. It will be difficult to distinguish actual scintillation from rapid variations of signal strength due to rain attenuation. In this chapter clear weather condition during a non rainy day is taken into account to study the scintillation effects on satellite communication signals. Tropospheric scintillation is characterised by the log-amplitude χ (in dB) and the scintillation variance σ^2_χ (in dB^2). The log-amplitude c is the ratio of the instantaneous amplitude of the observed signal to the mean amplitude in dB. The mean amplitude is calculated by averaging to 60 seconds (cutoff frequency around 0.01 Hz), in order to separate rapid turbulence-induced scintillation from spacecraft-induced variations. The scintillation variance σ^2_χ in dB^2 is the variance of the log-amplitude χ. In order to estimate the instantaneous scintillation effect considering its time-variability, the variance is normally calculated in a sliding way over one minute. A scintillation event is characterized by a stationary period of a few minutes (up to ten or fifteen minutes). Within this period, the scintillation variance remains approximately constant and the scintillation log-amplitude can be fitted quite well by a zero-mean Gaussian distribution whose variance is the mean scintillation variance during the period. For a period which is longer than several minutes, the variability of σ^2_χ is to be taken into account and the long-term probability density function of scintillation amplitude can be calculated by using Gaussian distribution with log-normally distributed variance.

$$\text{p long-term } (\chi) = \int_0^\infty p \text{ p short-term } (\chi/\sigma^2_\chi)\, p(\sigma^2_\chi)\, d\,\sigma^2_\chi \qquad \text{——(6.6)}$$

where p long-term (χ) is the long-term probability density function (pdf) of scintillation log amplitude and p short-term (χ/σ^2_χ) is the conditional short-term pdf of χ, assuming a constant value of σ^2_χ for the scintillation variance, which can be calculated from Gaussian distribution function

$$\text{p short-term } (\sigma_\chi^2) = 42.48 \frac{k^{7/6}}{(\sin\theta)^{11/6}} C_n^2(z) z^{5/6} \quad \text{——(6.7)}$$

p (σ^2_χ) is the long-term pdf of scintillation variance, which is log-normally distributed

$$p(\sigma_\chi^2) = \frac{1}{\sqrt{2\pi\sigma_\chi^2}\, s} \exp\left(\frac{-[\ln(\sigma_\chi^2 / m)]^2}{2s^2} \right) \quad \text{——(6.8)}$$

Where *m* (in dB²) and *s* (dimensionless) are the two parameters of the lognormal distribution. *m* is the median value of the scintillation variance distribution and *s* is the standard deviation of log (σ^2_χ).

Cumulative distributions of scintillation fade and variance, a common place way of representing long-term scintillation statistics, may be obtained by integrating (6-6) and (6-8), respectively. The scintillation variance cumulative distribution function can be derived from

$$cdf(\sigma_\chi^2) = \frac{1}{2} erfc\left(\frac{-\ln(\sigma_\chi^2 / m)}{\sqrt{2}s} \right) \quad \text{——(6.9)}$$

where cdf (σ^2_χ) is the scintillation variance cumulative distribution function or the probability that scintillation variance exceeds σ^2_χ and erfc ({1n (σ^2_χ/m)/$\sqrt{2}$s}) is the complementary error function.

The scintillation variance σ^2_χ can be related to the turbulence-induced refractive index structure parameter C^2_n in $m^{-2/3}$ and can be derived from

$$(\sigma_\chi^2) = 42.48 \frac{k^{7/6}}{(\sin\theta)^{11/6}} \int_{height} C_n^2(z) z^{5/6} \quad \text{——(6.10)}$$

where k = $2\pi/\lambda$, λ is the wavelength, θ is the elevation angle; z is the height above the ground level (in meters); and C^2_n (z) is the height dependent refractive index structure parameter (in $m^{-2/3}$).

Using the mean structure parameter profile extracted from balloon-borne soundings, the scintillations effect on a satellite link is estimated. The methodology to estimate C^2_n from radiosonde data is given below.

SCINTILLATION PREDICTION METHOD USING RADIOSONDE DATA

Radiosonde data is used to estimate the vertical profile of the refractive index structure parameter $C^2_n(z)$, which is related to the scintillation effect on earth-space paths by (6.10). The method comprises two main steps. First, representative statistical characteristics of the C^2_n profile are derived from an analysis of the available radiosonde data. C^2_n is a highly variable quantity whose values can be found between about 10^{-10} and 10^{-20} $m^{-2/3}$. There are mainly two regions in the troposphere in which turbulence is likely to be strong and C^2_n to be high: near the earth surface and in clouds. It has been demonstrated that turbulence in clouds (particularly fair weather cumulus) is responsible for most of the significant scintillation effects observed on satellite links.

The inference of fair estimates of the structure parameter profile from meteorological soundings is extensively dealt with in several papers. The method summarized hereafter is based on the work by *Warnock et al., 1985. Vaseur,H, 1999,* computed prediction formulas for tropospheric scintillation on satellite links by calculating height dependent refractive index structure parameter $C^2_n(z)$ using one year radiosonde data. According to the *Kolmogorov* theory, the radio frequency refractive-index structure parameter for well-developed turbulence is given by

$$C_n^2(Z) = a^2 \alpha' L_0^{4/3} M(Z)^2 \qquad \text{——(6.11)}$$

where a^2 is universal constant which is taken as 2.8 *(Ottersten, 1969)*, α^2 is the thermal to momentum diffusivity and varies slightly with atmospheric stability but it is taken as unity; L_0 is the outer scale length of the turbulence spectrum, its value being about 10 m and M is the vertical gradient of potential refractive index.

$$M(z) = -77.6 \times 10^{-6}\frac{P}{T}$$

$$\left[\frac{d\ln\theta}{dz} + 15{,}500\frac{q}{T}\left(\frac{d\ln\theta}{dz}\right) - \frac{7800}{T}\frac{dq}{dz}\right] \qquad \text{——(6.12)}$$

$$\theta = T\left(\frac{1000}{P}\right)^{0.286}$$

$$q = 0.622 \times \left(\frac{e}{P}\right)$$

$$e = 6.0178 \times \exp\left(5369 \times \left(\frac{1}{273.15} - \frac{1}{T(d)}\right)\right)$$

$$T(d) = \frac{[4880.357 - 29.66 \times \log(p_v)]}{[19.48 - \log(p_v)]}$$

$$p_v = \frac{R_h \times e_s}{100}$$

$$e_s = 6.112 \times \exp\left(6816 \times \left(\frac{1}{273.15} - \frac{1}{T}\right) + 5.1309 \times \log\left(\frac{273.15}{T}\right)\right)$$

if $T(d) = 273.0$ then $q = 0.0$;

if $T(d) \neq 273.0$ then $q = 0.622 \times \left(\frac{e}{P}\right)$

$$g(z) = 9.8 \times \left(\frac{6356776}{z + 6356776}\right)^2$$

where

R_h is Relative humidity (%)

T is Absolute temperature (K)

P is Atmospheric Pressure (hPa)

T(d) is Due point temperature (K)

θ is Potential temperature (K)

e_s is Saturated vapour pressure (hPa)

p_v is Vapour pressure (hPa)

q is Specific humidity (%)

e is water vapour pressure (hPa)

g is Acceleration due to gravity (m/s^2)

z is Altitude (km)

As the refractive-index gradient depends on meteorological parameters of the radiosonde data (values and gradients of pressure, temperature, and humidity), equation (6.11) is generally used to derive the structure parameter profile from the radiosonde data.

TROPOSPHERIC SCINTILLATION MONITORING TECHNIQUES

In this chapter scintillation variance σ^2_χ (in dB^2) is estimated for different elevation angles and for different frequencies, using the C^2_n values computed from the radiosonde data collected from Indian meteorological department observatory, Bangalore. Equations 6.10 and 6.11 are used for the computations. The estimated σ^2_χ (in dB^2) values are compared with observed scintillation variance over the earth satellite path of INSAT-3B. These estimates are useful for future planning. This type of work has not been reported earlier from this part of the world.

EXPERIMENTAL SETUP

In order to record the tropospheric scintillations of INSAT-3B satellite beacon signals of 11.544 GHz, the experimental set up shown in figure 4 is installed at Master Control Facility, Hassan, Karnataka. India. Hassan is located at 13.07° N, 76.08°E, 900 meters above mean sea level. To receive the INSAT-3B satellite signals the antenna bearings for Hassan are 151.78° Azimuth and 72.65° Elevation. The INSAT-3B satellite is located at 83° E over the equator. INSAT- 3B Indian coverage is shown in figure 3. Dish antenna of 3.8 m diameter is used to receive the beacon signal. The beacon signal is further down converted using LNA (Low Noise Amplifier) and fed to the beacon receiver. The output signal of beacon receiver is fed to the spectrum analyzer and to the computer for continuous recording of the satellite beacon signal at 5 seconds interval. The signal strength data was continuously recorded round the clock for four months from August 2004 to November 2004.

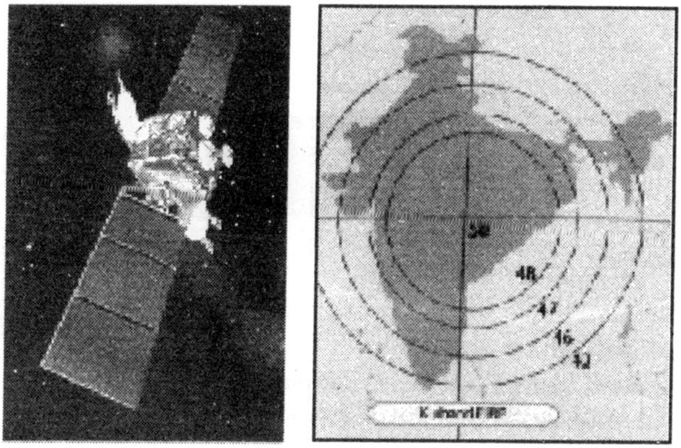

Figure 3: INSAT-3B Satellite & India Coverage

SCINTILLATION OBSERVATIONS

To analyze the tropospheric scintillation behavior and its effects over the satellite signals, observations were made for a 30 minute period in a non-rain day. Figures 5 to 7 show the signal zoom in plots for the days 10-08-04 (0800 to 0830 hrs.), 21-9-2004 (0930 t0 1000 hrs.) and 16-10-2004 (0800 to 0830 hrs.), chosen as non-rain days. From the signal strength data collected, daily graphs are plotted for particular days 10-8-2004, 21-9-2004 and 16-10-2004 as shown in figures 8 to 10 to analyse the signal variations. Tables 1 to 3 show the maximum, minimum signal levels, the peak to peak signal variations and standard deviation in two hours interval on 10-08-04, 21-9-2004 and 16-10-2004. From these tables, it can be seen that the peak-to-peak scintillation experienced is of the order of

2.12 dB on 21-09-2004 during 08-10 hrs while the minimum scintillation is experienced during 22-24 hrs at 0.279 dB on 10-08-2004. The standard deviation of this fast scintillation ranges between 0.04 to 0.2 dB.

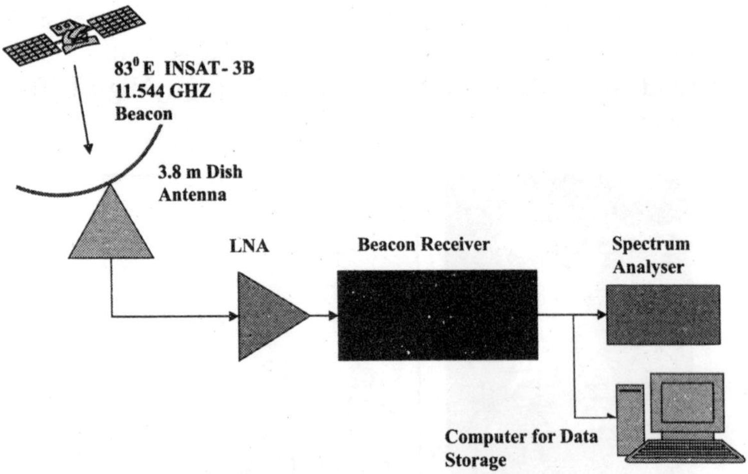

Figure 4: Block Diagram of Equipment Set up at Master Control Facility, Hassan, (Az. 151.78° & El. 72.65°)

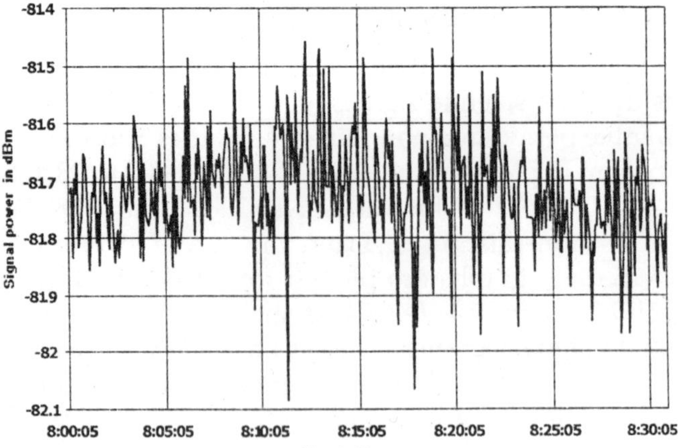

Figure 5: Signal Fluctuations on 10-08-2004 from 08:00 to 08:30 hrs

Table 1
Tables of Time Versus Scintillation for 10-08-2004

Time (Hr)	*Max. Scin.*	*Min. Scin.*	*Peak to Peak*	*Std. Dev.*
00-02	-81.4343	-81.8062	0.3315	0.04137
02-04	-81.4658	-81.8274	0.3616	0.05913
04-06	-81.4747	-82.0794	0.6047	0.09581
06-08	-81.6042	-82.1382	0.5340	0.08938
08-10	-81.4189	-82.0853	0.6664	0.08035
10-12	-81.3344	-81.9648	0.6304	0.08693
12-14	-81. 2016	-81.6768	0.4752	0.06978
14-16	-81.1764	-81.9648	0.7884	0.09216
16-18	-81.1941	-81.5513	0.3572	0.04702
18-20	-81.3226	-81.6101	0.2875	0.04740
20-22	-81.3616	-81.6600	0.2984	0.04316
22-24	-81.3962	-81.6753	0.2791	0.04151

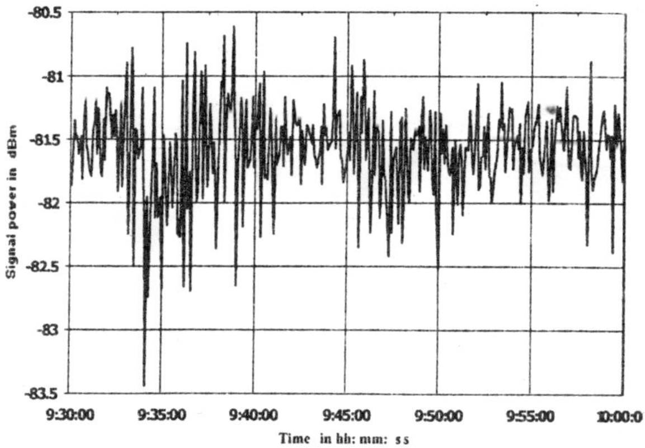

Figure 6: Signal Fluctuations on 21-09-2004 from 09:30 to 10:00 hrs.

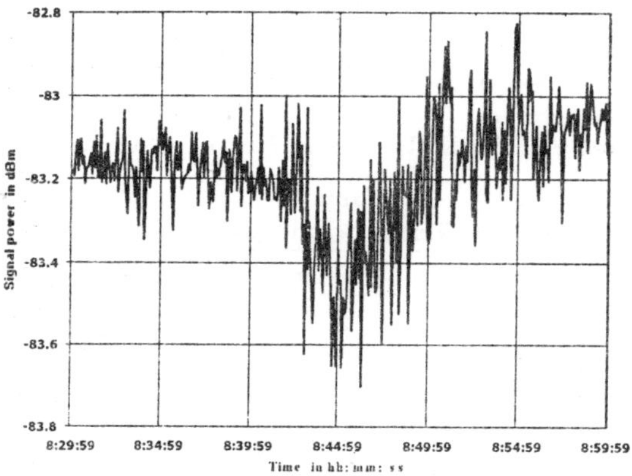

Figure 7: Signal Fluctuations on 16-10-2004 from 08:29 to 08:59 hrs.

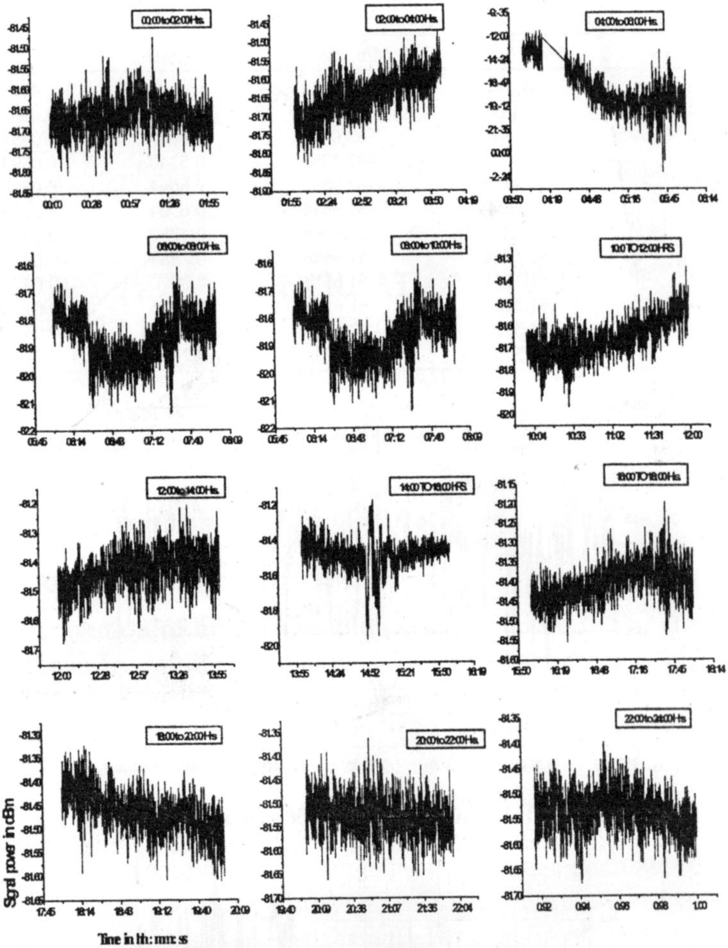

Figure 8: Signal Fluctuation Plot for 10-08-2004.

Table 2
Tables of Time Versus Scintillation for 21-09-2004

Time (Hr)	*Max. Scin.*	*Min. Scin.*	*Peak to Peak*	*Std. Dev.*
00-02	-81.4051	-81.8289	0.4238	0.05460
02-04	-81.4402	-81.8788	0.4386	0.05681
04-06	-81.4372	-82.0972	0.6600	0.09387
06-08	-81.5167	-82.4968	0.9801	0.11458
08-10	-80.6137	-82.7423	2.1231	0.2333

Contd.

10-12	-80.7194	-82.1046	1.3852	0.1577
12-14	-80.879	-81.4446	0.5656	0.09632
14-16	-80.8399	81.4248	0.5849	0.1287
16-18	-81.1699	-81.7776	0.6677	0.10116
18-20	-81.3359	-81.8062	0.4703	0.06877
20-22	-81.3903	-81.8215	0.4312	0.09193
22-24	-81.4021	-81.8062	0.4041	0.05464

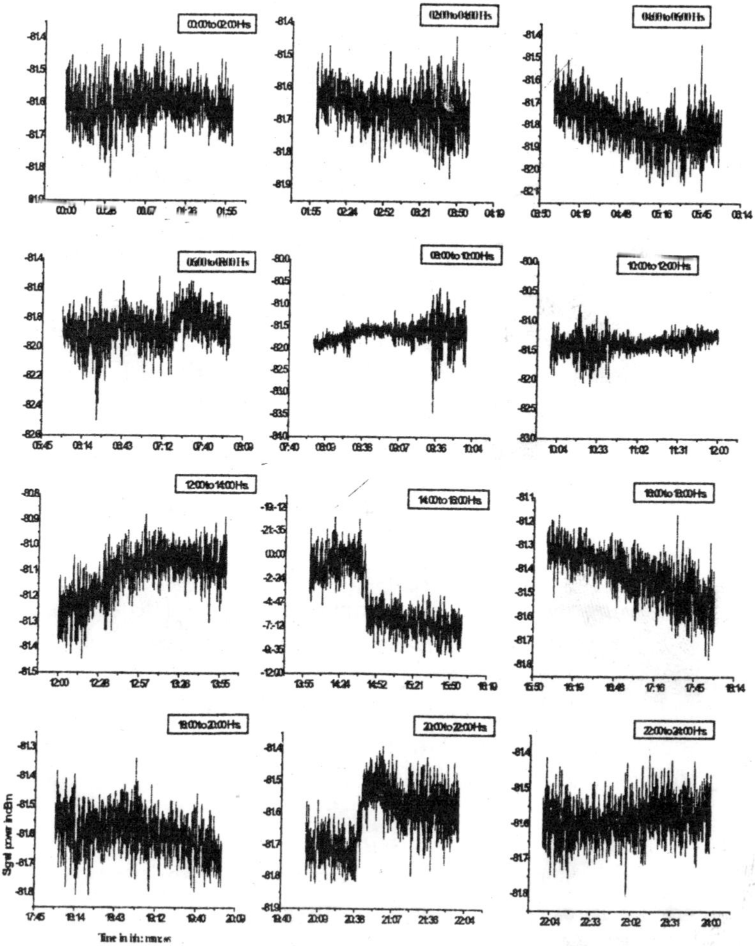

Figure 9: Signal Fluctuation Plot for 21-09-2004.

Table 3

Tables of Time Versus Scintillation for 16-10-2004

Time (Hr)	*Max. Scin.*	*Min. Scin.*	*Peak to Peak*	*Std. Dev.*
00-02	-83.5204	-83.9171	0.3967	0.07795
02-04	-83.4117	-83.7042	0.2925	0.05090
04-06	-83.2838	-83.7615	0.4777	0.07489
06-08	-82.8841	-83.7659	0.8818	0.13455
08-10	-82.8253	-83.7012	0.8759	0.09160
10-12	-82.7502	-83.267	0.5168	0.08998

Contd.

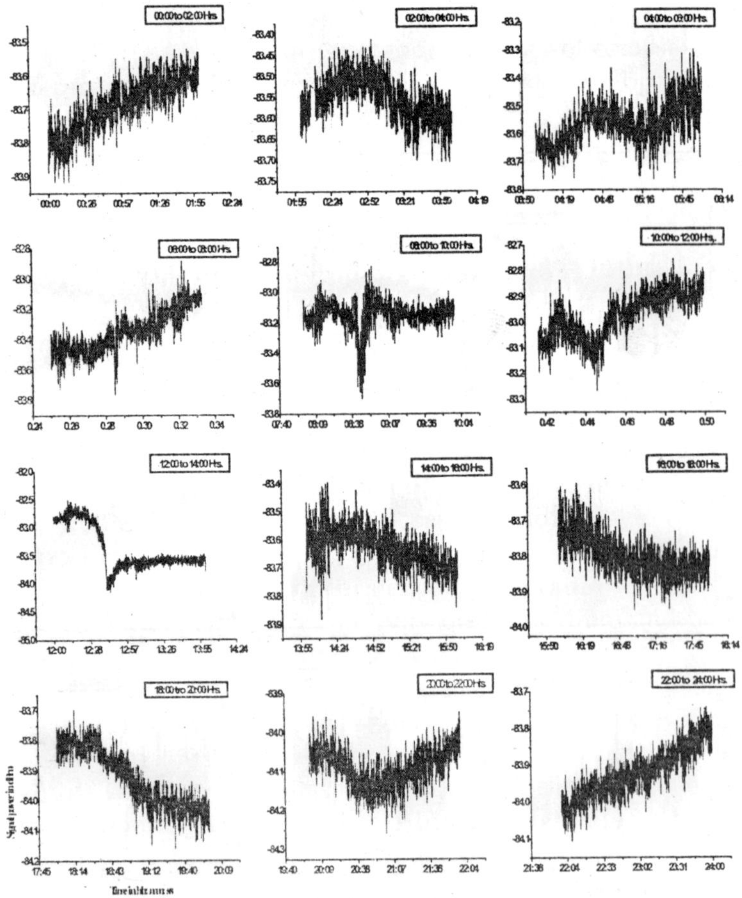

Figure 10: Signal Fluctuation Plot for 16-10-2004.

Time (Hr)	*Max. Scin.*	*Min. Scin.*	*Peak to Peak*	*Std. Dev.*
12-14	-82.4998	-84.1443	1.6445	0.33646
14-16	-83.3939	-83.8805	0.44866	0.06907
16-18	-83.593	-83.9531	0.3601	0.05706
18-20	-83.6967	-84.1581	0.4614	0.0981
20-22	-83.9472	-84.2589	0.3117	0.0494
22-24	-83.7422	-84.1053	0.3631	0.06819

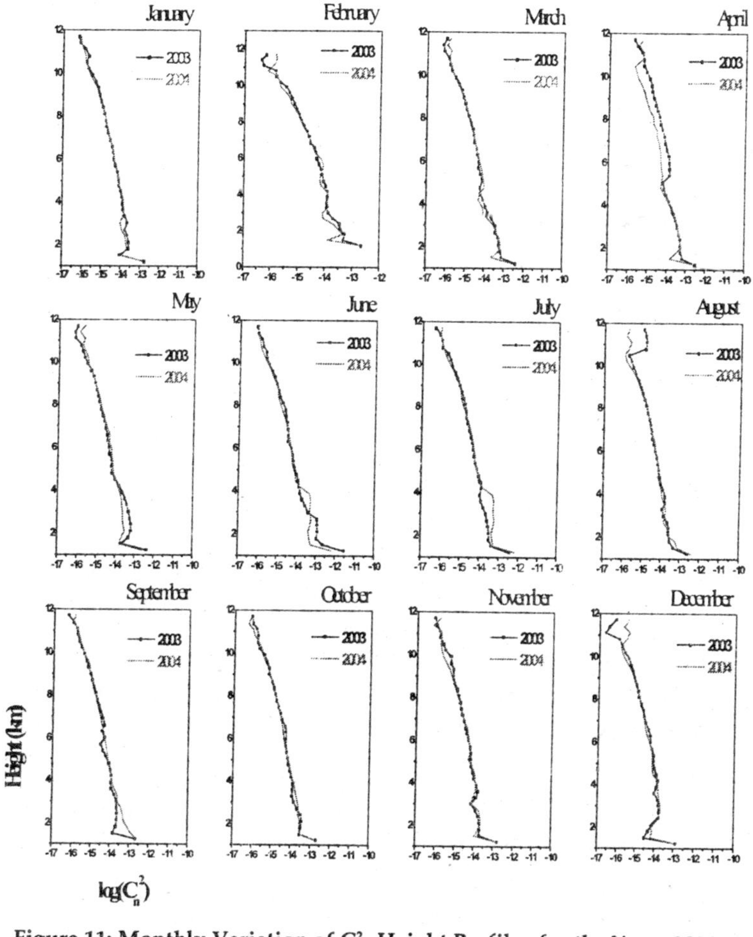

Figure 11: Monthly Variation of C_n^2 Height Profiles for the Years 2003 & 2004 at 0000 GMT over Bangalore, Computed Using Radiosonde Data

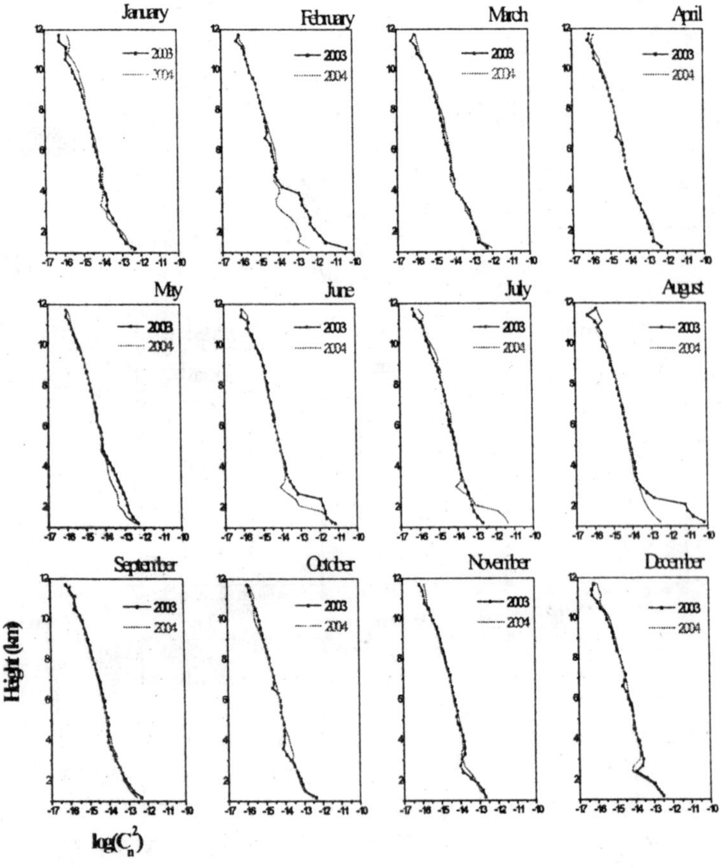

Figure 12: Monthly Variation of C^2_n Height Profiles for the Years 2003 & 2004 at 1200 GMT over Bangalore, Computed Using Radiosonde Data.

SCINTILLATION VARIANCE σ^2_χ (in dB^2) ESTIMATED FROM C^2_n VALUES FOR DIFFERENT ELEVATION ANGLES AND FREQUENCIES

In order to estimate σ^2_χ values, C^2_n is estimated from the radiosonde observations taken at Bangalore. Monthly variance of C^2_n height profiles for two years of observations (2003 & 2004) are shown in Figures 11 & 12. Figure 11 is for 0000 GMT and Figure 12 represents 1200 GMT. Log C^2_n values in general varies from -13 ($m^{-2/3}$) to -16 ($m^{-2/3}$). Higher values of C^2_n are

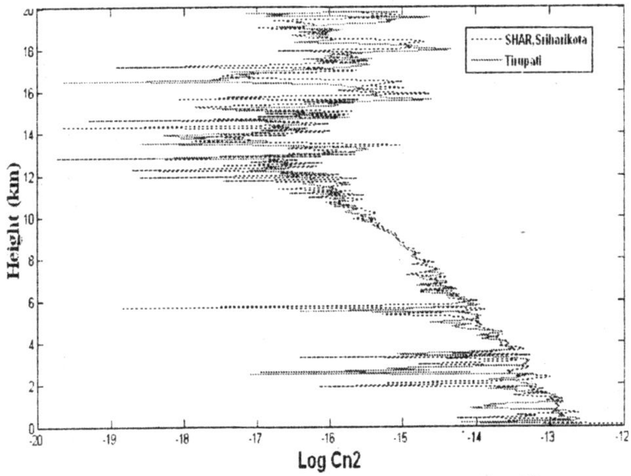

Figure 13: C^2_n Height Profiles Over Sriharikota and Tirupati , Computed Using GPS Sonde

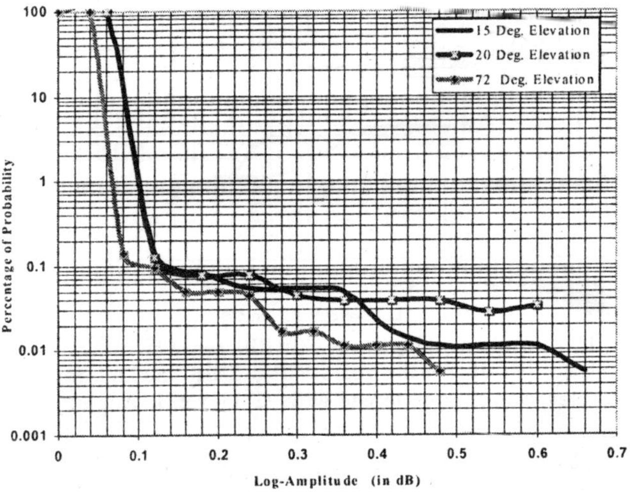

Figure 14: CDF of Log-Amplitude c (in dB) for Different Elevation Angles at 11.54 GHz for the Years 2003 & 2004.

observed at lower heights. No year to year variation is observed in C^2_n values in 0000 GMT. However, the C^2_n values estimated for 1200 GMT shows significant deviations in the monsoon (June-July-August-September) months. The large wind fluctuations results in the enhanced turbulence in the lower regions in monsoon months. The resolution of radiosonde is very poor, typically about 300 m and hence they can not resolve thin turbulent layers in the height profiles. GPS sonde data available at Tirupati and SHAR Centre is used to compute C^2_n values and are shown in figure 13. Fine turbulent layers of thickness ranging from 50 to 100 m can be seen from the figure. Log C^2_n values varies from about -13 ($m^{-2/3}$) to -16 ($m^{-2/3}$). This shows that there is no much spatial variation in the observed C^2_n values.

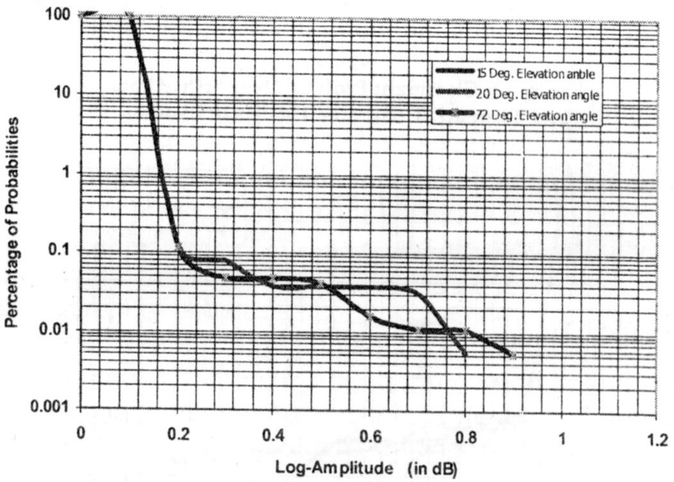

Figure 15: CDF of Log-Amplitude c (in dB) for Different Elevation Angles at 19 GHz for the Years 2003 & 2004. The Plot for 20° Elevation Angle Merged with 72° Elevation Angle.

From the variance σ^2_χ, the log-Amplitude χ (in dB) variations of scintillations were estimated for two years of C^2_n values computed from the radio sonde data of Bangalore. σ^2_χ values were calculated for 730 days at 0000 GMT and 730 days at 1200 GMT for 15°, 20° and 72° elevation angles for frequencies 11.54 GHz, 19 GHz and 30 GHz. The cumulative distribution functions are shown in Figs 14 to 16. From the figure 14 it can be observed that at low elevation angles the amplitude variations are higher and is about 0.6 dB at 0.01%

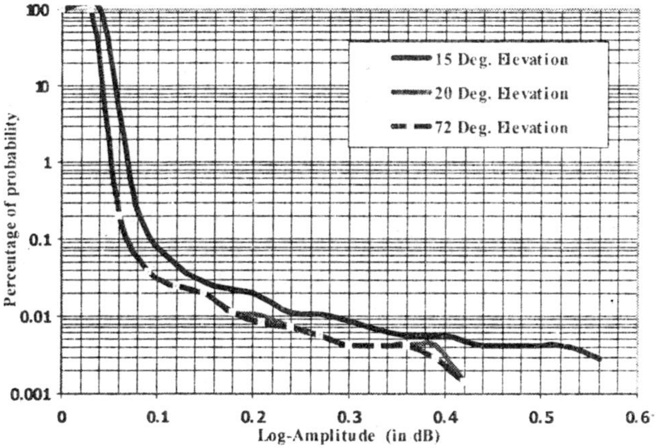

Figure 16: CDF of Log-Amplitude c (in dB) for Different Elevation Angles at 30 GHz for the Years 2003 & 2004.

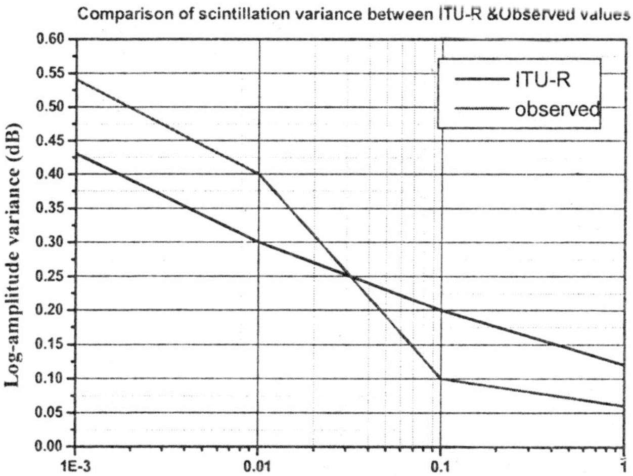

% of Probability

Figure 17: Comparison of Scintillation Variance Between ITU-R & Observed Values.

of probability. At 19 GHz (Figure 15) the observed scintillations are much higher compared to 11.54 GHz. However no difference is observed between 20^0 and 72^0 elevation angles. At 30 GHz (figure 16.) the variance is much less compared to 19 GHz frequency. The frequency dependence of scintillations are not very clear. *Van de Kamp et al., (1999)* have concluded from their measurements from many locations that there is no correlation between meteorological parameters and observed scintillations. A possible explanation provided by them is that scintillations due to cloudy turbulence has a different frequency dependence than 'dry' scintillations. Another reason could be due to the angle of arrival fluctuations have different frequency dependence and possibly play a role in turbulence-induced scintillations for highly directive antennas.

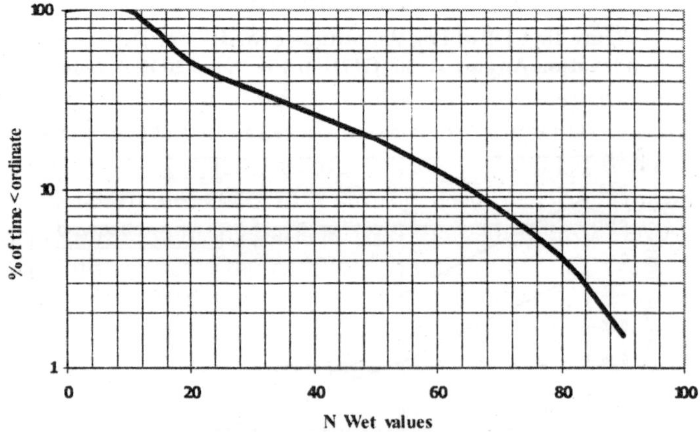

Figure 18: CDF of Wet Term of Refractivity, N for the Years 2003 & 2004.

Table 4
Results of ITU-R Model Run for Propagation Loss Computation for 1.0 % Probability

STATION		
Latitude	(degrés)	13.07
Longitude	(degrés)	76.08
Height	(km)	0.9
Elevation	(degrés)	72.65
ANTENNA		
Diameter	(m)	3.8
Efficiency	(%)	70

Contd.

LINK		
Frequency	(GHz)	11.544
Polarisation	(°)	45
Unavailability	(%)	1
CLIMATIC PARAMETERS		
R001	(mm/h)	3.62
Nwet	(N/A)	88.37
Rain height	(km)	5.08
Total Columnar Content Liquid Water	(kg/m²)	1.49
Surface Water vapour Density	(g/m³)	15.05
Integrated Water Vapour Content	(kg/m²)	45.61
Température	(°K)	296.36
ATTENUATION		
Atm. Gaseous Attenuation **(1)**	(dB)	0.07
Exceed Atm. Gaseous Attenuation **(2)**	(dB)	0.10
Rain Attenuation	(dB)	0.02
Clouds Attenuation	(dB)	0.19
Scintillation	(dB)	0.12
Total Attenuation	(dB)	(1)—0.32
		(1) & (2) —0.35

Table 5
Results of ITU-R Model Run for Propagation Loss Computation for 0.1 % Probability

STATION		
Latitude	(degrés)	13.07
Longitude	(degrés)	76.08
Height	(km)	0.9
Elevation	(degrés)	72.65
ANTENNA		
Diameter	(m)	3.8
Efficiency	(%)	70
LINK		
Frequency	(GHz)	11.544
Polarisation	(°)	45
Unavailability	(%)	0.1
CLIMATIC PARAMETERS		
R001	(mm/h)	31.37
Nwet	(N/A)	88.37
Rain height	(km)	5.08
Total Columnar Content Liquid Water	(kg/m²)	1.99
Surface Water vapour Density	(g/m³)	15.05
Integrated Water Vapour Content	(kg/m²)	49.74

Contd.

Température	(°K)	296.36
ATTENUATION		
Atm. Gaseous Attenuation **(1)**	(dB)	0.07
Exceed Atm.Gaseous Attenuation **(2)**	(dB)	0.11
Rain Attenuation	(dB)	1.74
Clouds Attenuation	(dB)	0.26
Scintillation	(dB)	0.20
Total Attenuation	(dB)	(1) —2.08
		(1) & (2)— 2.11

Table 6
Results of ITU-R Model Run for Propagation Loss Computation for 0.01 % Probability

STATION		
Latitude	(degrés)	13.07
Longitude	(degrés)	76.08
Height	(km)	0.9
Elevation	(degrés)	72.65
ANTENNA		
Diameter	(m)	3.8
Efficiency	(%)	70
LINK		
Frequency	(GHz)	11.544
Polarisation	(°)	45
Unavailability	(%)	0.01
CLIMATIC PARAMETERS		
R001	(mm/h)	90.82
Nwet	(N/A)	88.37
Rain height	(km)	5.08
Total Columnar Content Liquid Water	(kg/m^2)	2.31
Surface Water vapour Density	(g/m^3)	15.05
Integrated Water Vapour Content	(kg/m^2)	54.25
Température	(°K)	296.36
ATTENUATION		
Atm. Gaseous Attenuation **(1)**	(dB)	0.07
Exceed Atm. aseous Attenuation **(2)**	(dB)	0.11
Rain Attenuation	(dB)	10.49
Clouds Attenuation	(dB)	0.30
Scintillation	(dB)	0.30
Total Attenuation	(dB)	(1)—10.86
		(1) & (2) —10.90

The observed scintillations are compared with ITU-R model. The ITU-R results are shown in Tables 4 to 7. These tables are produced by running the on-line ITU-R model for different percentages of probabilities. Figure 17. shows the comparison between the observed and predicted values. It can be seen that ITU-R underestimates the variations, in the probability range 0.01 to 0.001 % where as the observed values are much lower in the probability range 1 to 0.1 %. The obvious reason for this could be with increasing temperatures, the wet part of refractivity steeply increases because for a given relative humidity, the water vapour pressure increases rapidly with increasing temperature. The wet term of N (88.37) in *ITU-R (Tables)* does not vary for different % of probabilities. The surface water vapour density (15.05 g/m^3) is also constant at all probabilities. The wet term refractivity is estimated for 2 years and its CDF is plotted and shown in figure 18. The wet term values are much higher than the ITU-R predicted values.

Table 7
Results of ITU-R Model Run for Propagation Loss Computation for 0.001 % Probability

STATION		
Latitude	(degrés)	13.07
Longitude	(degrés)	76.08
Height	(km)	0.9
Elevation	(degrés)	72.65
ANTENNA		
Diameter	(m)	3.8
Efficiency	(%)	70
LINK		
Frequency	(GHz)	11.544
Polarisation	(°)	45
Unavailability	(%)	0.001
CLIMATIC PARAMETERS		
R001	(mm/h)	154.78
Nwet	(N/A)	88.37
Rain height	(km)	5.08
Total Columnar Content Liquid Water	(kg/m^2)	2.69
Surface Water vapour Density	(g/m3)	15.05
Integrated Water Vapour Content	(kg/m^2)	59.16
Température	(°K)	296.36

Contd.

ATTENUATION		
Atm. Gaseous Attenuation **(1)**	(dB)	0.07
Exceed Atm. Gaseous Attenuation **(2)**	(dB)	0.12
Rain Attenuation	(dB)	23.16
Clouds Attenuation	(dB)	0.35
Scintillation	(dB)	0.43
Total Attenuation	(dB)	(1)-23.59 (1) & (2)-23.63

CONCLUSIONS

Tropospheric scintillations play a major role in earth-satellite paths especially at low elevation angles. It is believed that at higher elevation angles they are negligible. However the study presented in this chapter clearly demonstrated that the effect of scintillations can not be ignored even at high elevation angles. Using the methodology proposed by *Vasseur (1999)* log-amplitude variations of scintillations are estimated from C^2_n values. χ (dB) is estimated for different elevation angles and different frequencies. These values are very useful for planning of future earth-satellite paths in India. The estimated C^2_n values from radiosonde/GPS (Global Positioning System) sonde shows consistent results at Bangalore, SHAR Centre, Sriharikota and Tirupati. The ITU-R values under estimate the scintillation variance in the probability range 0.01 to 0.001.

Chapter–7

ATTENUATION OF INSAT 3B KU BAND SATELLITE COMMUNICATION SIGNALS DUE TO RAIN

INTRODUCTION

In recent years communication satellites are playing a major role in long distance communications. The advantage of the satellite era has ushered in high-tech space applications, the most important being telecommunications and broadcasting. The satellite industry has grown to handle the international telephone traffic, all international and domestic long distance television programs and rapidly growing domestic voice and data channels. The effects of precipitation on the transmission path are of major concern in space communications, particularly for those systems operating at frequencies above 10 GHz. Therefore, meteorological variations of humidity and rainfall need to be studied to predict propagation conditions and design effective receiving systems. The most fundamental obstacle encountered in design of satellite communication system at frequencies above 10 GHz is attenuation by rain. The increasing demand for these frequencies for telecommunications and broadcasting satellites has aroused increased interest in the study of radio wave attenuation due to rainfall on earth- satellite radio paths.

Rain is the main consideration, as the occurrence of other hydrometeors is very rare in tropical countries. The performance and reliability of communication systems, especially in tropical countries, are critically dependent on the attenuation caused by rain. The precipitation characteristics in the tropics differ appreciably from those of the temperate regions in that empirical relationships obtained in the latter may not be very suitable for system design in the former. Hence the current trend in employing the high microwave frequency bands for telecommunications and broadcasting services, necessitates the urgent need to carry out studies on the effect of rain on earth–space radio propagation, in tropics.

Rain attenuation can be obtained directly through measurement or predicted from a knowledge of rain rate and drop size distribution. The spectrum of raindrop sizes determines the rainfall rate, as well as the radar reflectivity and the microwave radiation emission of the raindrops. It is thus possible to estimate the attenuation by means of indirect measurements performed with rain gauges, radiometers and radar. The rain rates measured by rain gauges having short integration time are most relevant for prediction of attenuation of radio wave . The characteristics of the rainfall in a particular region must be investigated in the design of communications links through the region. High rain rates and the consequent high rain attenuation in the tropics is arguably the greatest constraint to the usability of the Ku and Ka bands in the tropics. The problem is further compounded by lack of reliable rain measurements and the present rain zones have not performed satisfactorily for attenuation calculations in the tropics.

In this chapter, preliminary studies on the Rain attenuation at 11.54 GHz is presented by monitoring INSAT-3B beacon signals from Hassan. Brief description of the monitoring techniques are presented in chapter-6 , while studying the effect of clear air scintillations on satellite paths. Brief description of rainfall characteristics and different rain attenuation prediction models are presented together with the attenuation results.

RAINFALL CHARACTERISTICS

Rainfall Structure

Rain is generally classified as Stratiform, Convective, Monsoon type and Tropical storms or Cyclones. Tropical rainfall is predominantly convective and characterized by high precipitation rates which occur, in general, over limited extensions and with short durations. Regardless of the precipitation type, rainfall is characterized by a space and time variable structure, constituted by cells of various dimensions that move horizontally, with speeds depending on the tropospheric winds and height of the clouds. The rain cell is formed by a nucleus with high precipitation intensity and a more extensive zone of lower intensity. The diameter of the nucleus is variable and inversely proportional to its intensity. Radar measurements have shown that typical dimensions of strong rain rate cells range from 2 to 5 km.

Stratiform Rain

Rainfall of this nature normally occurs in non-tropical regions and is characterized by low rain rates with small embedded showers and horizontal extent of hundreds of kilometers.

Convective Rain

This type of rain occurs due to vertical motion of an ascending mass of air which is warmer than its environment. These occur when the rain drops which form at the bottom of the cloud are transported by the up drafts to a height where their concentration increases quickly and then fall back to the ground due to violent down drafts *[(Moupfouma, 1987(b)]*. It is characterized by high fall rates with duration of tens of minutes and a horizontal scale of several kms. In equatorial climates this type of rainfall gives rise to more thunderstorms than elsewhere.

Monsoon Rain

This is due to intense convective rainfall followed by intervals of statiform rain which constitute monsoon rains. Heavy precipitation is observed lasting for several hours across hundreds of km in this kind of rain fall.

Tropical Cyclones

In tropics this type of rainfall is experienced with intense convective rain of 50-200 km in diameter and extending to 8 km or so vertically. The storms are characterized by several spiral bands terminating in regions of intense precipitation and lasts for several hours of the day.

RAINFALL RATES

Seasonal and year-to-year variation of the rainfall rate distribution occurs due to normal and time varying characteristic of the rainfall. Due to rapid variation of rainfall at a given point, the cumulative rainfall rate distribution measured will be dependent on the sampling time of the rain gauge. Rain gauges with short integration time are used for rainfall rate measurement and this data is used for the prediction of attenuation of radio wave *(Watson et al., 1987)*. The tipping bucket rain gauge is the most commonly used rain guage. Intense rain can saturate this type. The Siphon rain gauge is too slow in reaction for use in areas of tropical rain because they do not indicate sharp intensity peaks. Fast response rain gauges are being used recently with the capability of measuring high rain rates of 250 mm/hr. Rain rate is required for short and intense bursts occurring for 0.01 to 0.001 percent of time and for this type of data, rain gauges of short integration times of few seconds, are needed.

DISTRIBUTION OF DROP SIZE

The rain structure constitutes of six parameters. They are (i) Rain rate. (ii) Liquid water content, (iii) Median drop diameter, (iv) specific attenuation, (v) radar reflectivity factor and (vi) optical extinction *(Ulbrich, 1983)*. Recording rainfall rate and drop size distribution is an easy and inexpensive method compared to attenuation measurement. Different models are used for attenuation computations. The different models for drop size distribution measurement are *Laws and*

Parsons (1943), Marshall and Palmer (1948), Muller and Jones (1960), Joss et al., (1968), Barelay et al., (1978), Moupfouma (1984) and Ajayi and Olsen (1985). All the first five models are developed based on temperate climatic data where as the last two models are developed on the basis of tropical measurements.

EFFECTIVE RAIN HEIGHT

For calculation of slant path attenuation, the rainfall height is an important parameter. It is generally considered that the rain structure reaches a maximum height equal to the 0° isotherm. Above that height, precipitation is assumed to have the form of ice, snow or melting snow. The importance of this freezing height is due to the fact that the radio wave attenuation is the largest in the liquid form of hydrometers (dielectric constant of liquid water is 80 times that of free space) and much lesser in frozen forms. However the tropical rainstorms, which are of convective nature, reach altitudes of 6 to 10 km, far above the 0° isotherm. For vertical attenuation predictions, the height of 0° isotherm depends on different meterological conditions. It is found *(Oort and Rasmusson, 1971)* that the average height of 0° isotherm varies from 4.7 km in the tropics to 3.1 km at latitudes 40^0-60^0. The concept of effective rain height was developed in the tropical regions due to large deviations obtained when prediction models are applied to some regions. The effective rain height (h_{FR}) was equated to the 0° isotherm height during rainy conditions and h_{FR} *(ITU-R p.839-3, 2001)*, can be obtained from radiosonde data or other sources. The vertical profile of rain and its height in particular are taken into consideration by the ITU-R in the prediction of rain attenuation along earth satellite paths.

ESTIMATION OF SPECIFIC ATTENUATION

Specific attenuation calculation for a known frequency for different rain rates strongly depends on the assumed microstructure of rainfall; size distribution, temperature, terminal velocity and shape of raindrops. The specific attenuation g in dB/km due to precipitation can be evaluated on the basis of classical Mie scattering theory. The relationship between specific attenuation g (dB/km) and rain rate (mm/hr) for practical applications is approximated by the *power law (Olsen et al., 1978).*

$$\gamma = kR \qquad \text{—(7.1)}$$

where k and a are parameters depending on wave frequency and polarization, for a given model of the rain microstructure. Values of k and a for the frequency range from 1 to 400 GHz were calculated *(Fedi, 1979; Maggiori, 1981)* for raindrops with the assumed shape of oblate spheroids, at the temperature of 20⁰ C, using the drop size distribution proposed by *Laws and Parsons (1943)*, terminal velocity following *Gunn and Kinzer (1949)* and refractive index values according to the model of *Ray (1972). ITU-R* has recently revised their recommendations No.P.838-2 and the frequency dependent coefficients for estimating specific attenuation are shown in Table 1, for frequencies between 1 and 400 GHz, for vertical and horizontal polarizations.

ATTENUATION PREDICTION MODELS

Rain attenuation can be obtained directly through measurement or can be predicted from the data of rain rate, drop size distribution and other relevant parameters of radio path. Various models are available for the prediction of attenuation due to rain on earth-space path. These models are used for the estimation of rain attenuation, in general where ever adequate data is not available. Various prediction models such as ITU-R recommendation *P.618-8 (2003). Garcia-Lopez (1988), Moupfouma's [1987 (a)],DAH model [Dissanayake et.al., 1997] Crane two component model [1999]* proposed a new approach for modeling rain attenuation on earth-satellite links and LOS links. The following models are also considered by ITU-R in their study: The Brazil model (1992); The Japan model (Karasawa, 1989); U.K. model (Leitao-Watson, 1986); France Model (Misme-Waldtenfel, 1980) and Excell Model (Capsoni, et.al., 1987).

Table 1
Frequency Dependent Coefficients for Estimating Specific Attenuation As Per Recommendations of ITU-R P.838-2

Frequency(GHz)	K_H	k_V		
1	0.0000387	0.0000352	0.9122	0.8801
1.5	0.0000868	0.0000784	0.9341	0.8905
2	0.0001543	0.0001388	0.9629	0.9230
2.5	0.0002416	0.0002169	0.9873	0.9594
3	0.0003504	0.0003145	1.0185	0.9927
4	0.0006479	0.0005807	1.1212	1.0749
5	0.001103	0.0009829	1.2338	1.1805
6	0.001813	0.001603	1.3068	1.2662
7	0.002915	0.002560	1.3334	1.3086
8	0.004567	0.003996	1.3275	1.3129
9	0.006916	0.006056	1.3044	1.2937
10	0.01006	0.008853	1.2747	1.2636
12	0.01882	0.01680	1.2168	1.1994
15	0.03689	0.03362	1.1549	1.1275
20	0.07504	0.06898	1.0995	1.0663
25	0.1237	0.1125	1.0604	1.0308
30	0.1864	0.1673	1.0202	0.9974
35	0.2632	0.2341	0.9789	0.9630
40	0.3504	0.3104	0.9394	0.9293
45	0.4426	0.3922	0.9040	0.8981
50	0.5346	0.4755	0.8735	0.8705
60	0.7039	0.6347	0.8266	0.8263

Contd.

Frequency(GHz)	K_H	k_V		
70	0.8440	0.7735	0.7943	0.7948
80	0.9552	0.8888	0.7719	0.7723
90	1.0432	0.9832	0.7557	0.7558
100	1.1142	1.0603	0.7434	0.7434
120	1.2218	1.1766	0.7255	0.7257
150	1.3293	1.2886	0.7080	0.7091
200	1.4126	1.3764	0.6930	0.6948
300	1.3737	1.3665	0.6862	0.6869
400	1.3163	1.3059	0.6840	0.6849

The ITU-R model has the objective of being globally applicable across a wide range of frequencies, elevation angles, and rain climates. The DAH model seeks to improve upon the overall ITU-R model performance by modifying path profiles as well as adjusting the calculations across a wider range of availabilities. Both the Japan model and Brazil model are developed as refinements to the ITU-R model which focus on improving prediction accuracy at lower system availability levels. The Brazil model adds the additional refinement of increasing accuracy for systems operating in tropical and/or equatorial regions.

The remaining models are independently derived and are primarily based on radar data. Misme and Waldteufel's model is an extension of their terrestrial rain attenuation model which considers rain to consist of a circular cell with uniform rain rate and "a weakly rainy area" surrounding the cells. The Two Component model also proposes a technique for individually calculating the effects of isolated cells and the surrounding "debris". The two sources of rain attenuation are derived independently and then summed. The Leitao-Watson model attempts a "first principals" approach which extrapolates point to path characteristics from dual polarization radar cross section data and applies that data to scattering theory. The extension to Earth-space paths is essentially a geometric exercise. ExCell is an abbreviation for Exponential CELL model. This model, which is also derived from radar measurements, takes an empirical approach to modeling a typical cell structure. Multiple cells are then used to recreate the total distribution of rain rates occurring across a path. Garcia-Lopez provides a technique which is optimized for simplicity of calculations and maximum accuracy at the highest attenuation values. The working party 3M of the ITU-R presented research comparing the estimations of rain attenuation models with the 186 station years of earth-space propagation data. It was concluded that the overall best performing model is the DAH, based on the ITU-model.(Glenn S.Feldhake etal., 2002). Hence, in this chapter, for the rain attenuation prediction models *ITU-R.P.618-8*

[2003], Garcia-Lopez [1988], Moupfouma [1987 (a)] and Dissanayake- Allnutt- Haidara (DAH) model [Dissanayake et al.,1997] are only used to calculate the effect of rain on Ku band satellite signals the results are compared with the observed values. Description of these models is given below.

ITU-R's Method

This method enables estimation of the long-term statistics of the slant-path rain attenuation at a given location for frequencies up to 55 GHz, using the following parameters and the formulas mentioned below:

- $R_{0.01}$: point rainfall rate for the location for 0.01% of an average year (mm/h)
- h_s : height above mean sea level of the earth station (km)
- θ : elevation angle
- φ : latitude of the earth station (degrees)
- f : frequency (GHz)
- R_e : Effective radius of the Earth (8500km)

The effective rain height measurement is shown in figure 1.

Formula 1: To calculate the effective rain height, h_R, for the latitude of the station φ :

The Rain height, h_R as given in Recommendation ITU-R p.839-3(2001)

$$h_R = h_0 + 0.36 \qquad \text{—(7.2)}$$

where h_0, yearly average 0°C isotherm height above msl(km)

Formula 2: For $\theta \geq 5^0$ from the slant-path length, L_s, height below rain $(h_R - h_s)$ can be calculated from :

$$L_s = \frac{(h_R - h_s)}{\sin\theta}\,(\text{km}) \qquad \text{—(7.3)}$$

For $\theta < 5^\circ$, the following formula is used:

$$L_s = \frac{2(h_R - h_s)}{\left[\sin^2\theta + \frac{2(h_R - h_s)}{R_e}\right]^{\frac{1}{2}} + \sin\theta} \text{ (km)} \qquad \text{—(7.4)}$$

Formula 3: To calculate the horizontal projection, L_G of the slant path length from:

$$L_G = L_s \cos\theta \text{ (km)} \qquad \text{—(7.5)}$$

Formula 4: To obtain the rain intensity, $R_{0.01}$, exceeding for 0.01% of an average year (with an integration time of 1 min).

If this information cannot be obtained from local data sources, an estimate can be obtained from the maps of rain climates given in *Recommendation ITU-R P.837.*

Formula 5: To obtain the specific attenuation, γ_K using the frequency-dependent coefficients given in Recommendation ITU-R P.838 and the rainfall rate, $R_{0.01}$, determined from Step 4, by using:

$$\gamma_R = k(R_{0.01})^a \text{ (dB/km)} \qquad \text{—(7.6)}$$

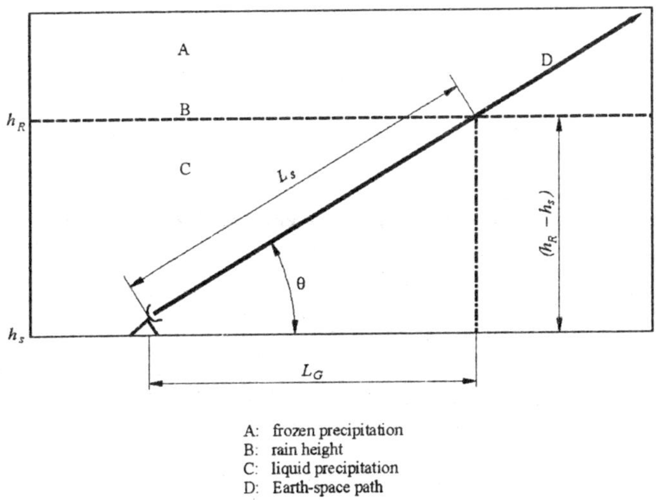

Figure 1: The Effective Rain Height Measurement

Formula 6: To calculate the reduction factor, $r_{0.01}$, for 0.01% of the time

$$\nu_{0.01} = \frac{1}{1 + 0.78\sqrt{\frac{L_G \upsilon_R}{f}} - 0.38\left(1 - e^{-2L_G}\right)} \quad \text{---(7.7)}$$

Formula 7: Calculate the vertical adjustment factor, $u_{0.01}$, for 0.01% of the time:

$$\xi = \tan^{-1}\left(\frac{h_R - h_s}{L_G r_{0.01}}\right) \text{ degrees}$$

For $\xi > \theta$, $L_R = \frac{L_G r_{0.01}}{\cos\theta}$ km

Else, $L_R = \frac{(h_R - h_s)}{\sin\theta}$ km

If $\varphi < 36°$, $\chi = 36 - \varphi$ degrees
Else, $\chi = 0$ degrees

$$\nu_{0.01} = \frac{1}{1 + \sqrt{\sin\theta}\left(31\left(1 - e^{-(\theta/(1+\chi))}\right)\frac{\sqrt{L_R \gamma_R}}{f^2} - 0.45\right)}$$

The effective path length is:

$$L_E = L_R V_{0.01} \text{ km}$$

The predicted attenuation exceeded for 0.01% of an average year is obtained from:

$$A_{0.01} = g_R L_E \text{ dB}$$

The estimated attenuation to be exceeded for other percentages of an average year, in the range 0.001% to 5%, is determined from the attenuation to be exceeded for 0.01% for an average year:

If $p \geq 1\%$ or $\varphi \geq 36°$: $\beta = 0$

If $p < 1\%$ and $\varphi < 36°$ and $\theta \geq 25°$: $\beta = -0.005\ (\varphi - 36)$

Otherwise:

$$\beta = -0.005(\varphi - 36) + 1.8 - 4.25 \sin\theta$$

$$A_p = A_{0.01}\left(\frac{p}{0.01}\right)^{-(0.655 + 0.033\ln(p) - 0.045\ln(A_{0.01}) - \beta(1-p)\sin\theta)} \text{dB}$$

Garcia-Lopez Method

According to this method, the attenuation A is given by

$$A = kR^{\alpha} L_s \div \left\{ a + \left[\frac{L_s(bR + cL_s + d)}{e} \right] \right\} \qquad \text{—(7.10)}$$

k, and α are obtained from *ITU-R Recommendations p.838-2,* R is the rain rate in mm/hr, L_s is the equivalent path length in km, a, b, c, d and e are constants . Two sets of values for a, b, c, d and e, one set for worldwide use and the other for Australia. The values mentioned for Australia can be used for tropical climatic conditions

(a=0.72, b = 76, c=-4.75, d= 2408 and e = 10,000)

Moupfouma's Method

Values of the elevation angle of the slant path, height of the station above sea level etc., are needed for this method. The concept of correction factor introduced in this method requires the rain rate of 0.01 percent of time and rain rates for other time percentages. The predicted attenuation is given by

$$A = \frac{kR^{\alpha} u(p)(R_{0.01}/R_p)0.38L_s^{-0.25}}{1 + \eta(p/0.01)^{-0.36} L_s^m} \times L_s \qquad \text{—(7.11)}$$

where $m = 1 + 1.4 \times 10^{-4} f^{1.76} \log_e(L_s)$, f is frequency in GHz, L_s is slant path length (km)

$\eta = 0$ if $L_s < 5$ km, $f < 25$ GHz

$= 0.03$ in all other cases.

The values of parameter *'u'* for different regions of the world are given in a separate table by *Moupfouma.* $R_{0.01}$ and R_p are the rain fall rates corresponding to rain rates of 0.01 and p percent of times.

The slant path length L_s is given by

$$L_s = \frac{H_R - H_0}{\left[\sin^2\theta + 2\left(\frac{H_R - H_0}{8500}\right)^{1/2}\right] + \sin\theta} \quad \text{—(7.12)}$$

θ is the elevation angle, H_0 is the station height above mean sea level. H_R is the height of 0°C isotherm.

Dissanayake-Allnutt-Haidara (DAH) Method

In this method, inhomogeniety of rain in horizontal and vertical directions are accounted for prediction and is applicable to the frequency range 4 -35 GHz. This model is similar to ITU-R rain attenuation prediction method.

The Input Data Required for the Model are:

- ϕ : latitude of the earth station (degrees)
- h_s : altitude of the earth station above mean sea level (km)
- *R 0.01* : point rain fall rate for the location for 0.01% of an average year (mm/h)
- p : percentage exceedance probability for which attenuation is to be calculated.
- θ : elevation angle
- f : frequency (GHz)
- ξ : Polarisation angle (degrees)
- R_e : effective earth radius (8500 km)

The following Formulas are used.

Formula 1: Freezing height during rain h_{fr} (km) is calculated from the absolute value of the station latitude ϕ (degrees):

$$h_{fr} = 5.0 \text{ for } 0^0 \le \phi \le 23^0$$

$$h_{fr} = 5.0 - 0.075\,(\phi - 23),\ \text{for } \phi\ 23^0 \quad \text{—(7.13)}$$

Formula 2: Slant path L_s, below the freezing height is

$$L_s = \frac{h_{fr} - h_s}{\sin\theta}\ (\text{km}) \quad \text{—(7.14)}$$

For $\theta < 5^0$, the following formula is used:

$$L_s = \frac{2(h_{fr} - h_s)}{\left[\sin^2\theta + 2\left(\frac{h_R - h_S}{R_e}\right)\right]^{1/2} + \sin\theta}\ (\text{km}) \quad \text{— —(7.15)}$$

Formula 3: The horizontal projection L_g of the slant path length is calculated from:

$$L_g = L_s \cos\theta\ \ (\text{km}) \quad \text{— (7.16)}$$

Formul 4: From the rain intensity, $R_{0.01}$ (mm/hr), for 0.01% of an average year and calculate the specific attenuation (dB/km) can be calculated from

$$\gamma = K(R_{0.01})^{\alpha} \quad \text{—(7.17)}$$

Formula.5: The horizontal path adjustment factor, $r_{0.01}$, for 0.01% of the time is calculated from

$$r_{h0.01} = \frac{1}{1 + 0.78\sqrt{\frac{L_g\gamma}{f}} - 0.38\left[1 - \exp(-2L_g)\right]} \quad \text{—(7.18)}$$

Where f is the frequency in GHz.

Formula.6: The adjusted rainy path length L_r (km) through rain is calculated from

$$L_r = \frac{L_g rh_{0.01}}{\cos\theta}\ \ \text{for } Z > 0 \quad \text{—(7.19)}$$

$$L_r = \frac{(h_{fr} - h_s)}{\sin\theta} \; for \xi \leq \theta$$

Formula.7: The vertical adjustment factor $rv_{0.01}$ for 0.01% of time is calculated from

$$r_{v0.01} = \frac{1}{1 + \sqrt{\sin\theta}\left[31\left(1 - e^{-\theta[1+\chi]}\right)\sqrt{\frac{L_r\gamma}{f^2}} - 0.45\right]} \qquad \text{—(7.20)}$$

where

$$\chi = 36 - |\phi|, for |\phi| < 36^0 \, Latitude$$

$$\chi = 0, for |\phi| \geq 36^0 \, Latitude$$

Formula. 8: The effective path length through rain L_e (km) is calculated from

$$L_e = L_r r_{v0.01} \text{ (km)} \qquad \text{—(7.21)}$$

Formula.9: The attenuation exceeded for 0.01% of an average year may then be obtained from

$$A_{0.01} = \gamma L_e \text{(dB)} \qquad \text{—(7.22)}$$

Formula 10: To estimate the attenuation exceeded for other percentages of an average year, in the range 0.001% to 10% is determined from the attenuation to be exceeded for 0.01% for an average year by using:

$$A_p = A_{0.01}\left[\frac{p}{0.01}\right]^{-[0.655 + 0.033 \log p - 0.045(nA0.01) - Z \sin\theta(1-p)]} \qquad \text{—(7.23)}$$

Where p is the percentage probability of interest and Z is given by

For $p \geq 1\% Z = 0$

For p< 1%

$Z = 0, for |\phi| < 36^0$

$$Z = -0.005\left(|\phi| - 36\right), for\theta \geq 25^0 and |\phi| < 36^0$$

$$Z = -0.005\left(|\phi| - 36\right) + 1.8 - 4.25\sin\theta, for\theta < 25^0 and |\phi| < 36^0$$

WORK DONE SO FAR

Crane,(1980); Allnutt,(1989); Crane, (1996); Matricciani, (1997) (a), (b):Crane,(2003); Matricciani, (2004) made considerable research work on rain attenuation measurements of earth-space paths of Geostationary satellites, namely, the NASA Applications Technology Satellite (ATS), the Canadian/US Communications Technology Satellite (CTS), the domestic US COMSTAR communications satellite, the ETS,CS and BSE of Japan, the Italian SIRIO and OTS of Europian Space Agency, INSAT and MEASAT. *Hogg and Chu (1975)* also analysed rain attenuation data from communication satellites. *Davies and Mackenzie (1981)* conducted SHF/EHF slant path propagation measurements made at Slough, UK and Brussard (1981) reviewed the prediction techniques of slant path attenuation. *Nackoney and Davidson (1982) and Oguchi (1983)* have conducted satellite-earth rain attenuation at 11.7 GHz, utilizing CTS beacon signals. *Crane (1985)* has proposed rain attenuation prediction models which are based on point rain rate distribution. *Capsoni (1987)* had evolved a method for the prediction of rain attenuation at frequencies above 10 GHz. *Vogel et al., (1993)* conducted rain fade measurements at low elevation angles at 11.2 GHz. *Matricciani (1998)* carried out measurements of rain attenuation at 11.6 GHz. The author is also involved earlier on rain attenuation studies at 11.7 GHz and the results were published *(S.V.B.Rao et at.,)*

Several research works were carried out on slant path attenuation using the point rainfall statistics *[Manning,1984; Moupfouma, 1984; Ajaji, 1984]. Dissanayake et al., (1990)* conducted radiometric rain attenuation measurements at 11.6 GHz in Peru. *Pontes et al. (1990)of Brazil, McCarthy et al. [1992 (a), (b), (c)]* (Cameroon, Kenya and Nigeria) *Brussad et al. (1993) of Indonesia, Lekkla et al., (1994)* of south –East Asia, *Rodda and Willamson (1996)* of Auckland, New Zealand, carried out slant path radiometric measurements at 11.6 GHz. *Adji et al., (1995)* conducted rain attenuation studies at 11.2 GHz for the Indonesian satellite. *Polonio and Riva (1998)* carried out on ITELSAT satellite propagation measurements for the frequency bands 20, 40, 50 GHz on attenuation, scintillation and fade duration. *Yamada and Miura (1998)* carried out rain attenuation characteristics at 12 GHz on an earth-space path using Japanese Superbird-B satellite. Recently *Van de Kamp (1999)* has proposed fade countermeasure "Adaptive Power Control" to reduce communication outage due to rain fading.

Significant work on slant path attenuation was reported over Singapore *(Ong, J.T. et.al., 2000; Timothy et.al., 2000;2001).* A comprehensive fade dynamics review was provided by *van de Kamp et.al., (2002).* Global mapping of Attenuation at Ku and Ka band was done by *Chandrasekhar (2003)* using TRMM data. *Ramachandran et.al., (2003)* have reported attenuation studies over Fizi island. Site diversity characteristics were studied by *Takeshi Hatsuda (2004).* Spatio-Temporal rain attenuation model for application to fade mitigation techniques was developed by *Gremont et.al., (2004).*

Various prediction models of rain attenuation in satellite communication links are developed mainly on measurements carried out in temperate climate zones. In tropical areas like India systematic measurements were not carried out in Ku band. The data available in India is from radiometer observations over New Delhi (*Raina and Uppal, 1981)* and are not sufficient to predict rain attenuation in satellite communications. In India significant work was reported on rain attenuation on microwave communications (Timothy et.al.,1995,1997(a),(b);1998); Sarkar et.al., 2003;2004;2005). Rain attenuation studies over Terrestrial mocrowave link at 19.4 GHz was reported by Ashok Kumar, (2004). Ashok Kumar (2004) has also reported Zenith-path attenuation at 19.9GHz using radiometer at Amritsar. Extensive work on rain height, 0°C isotherm was reported from National Physical Laboratory, New Delhi (Sarkar et.al., 1998; 2003; Mondal et.al., 2001. Cloud height, cloud temperature and cloud attenuation over Indian Tropical east coast was reported by Sarkar et.al., (2005). However most of the above studies are confined to Northern India and North-east of India. In Southern India no significant work was reported.

EXPERIMENTAL SETUP

As a part of this work, rain attenuation studies were conducted in southern India at INSAT Master Control Facility, Hassan (76.08^0 East, 13.07^0 North, 900 meters above mean sea level), Karnataka, India. INSAT-3B Satellite located at 83.5^0 E longitude. Its beacon frequency is continuously recorded for 5 seconds interval on 24 hour basis for four months from August 2004 to November 2004 using 3.8 meter dish antenna. Rain rate data is collected using rapid response rain guage (Dynalab, make) with 5 second integration at the same place where beacon signal is monitored. The beacon signal output from the beacon receiver and the rain rate sensor output are fed to the Sefram Strip chart Recorder and Rohde Schwarz Spectrum Analyser. Both the signal strength and the rain rate are simultaneously recorded for every 5 seconds and the data collected for further analysis. The experimental setup for data collection is shown in Figure 2. Efforts were made to record signal strength fadings during heavy rainy days.

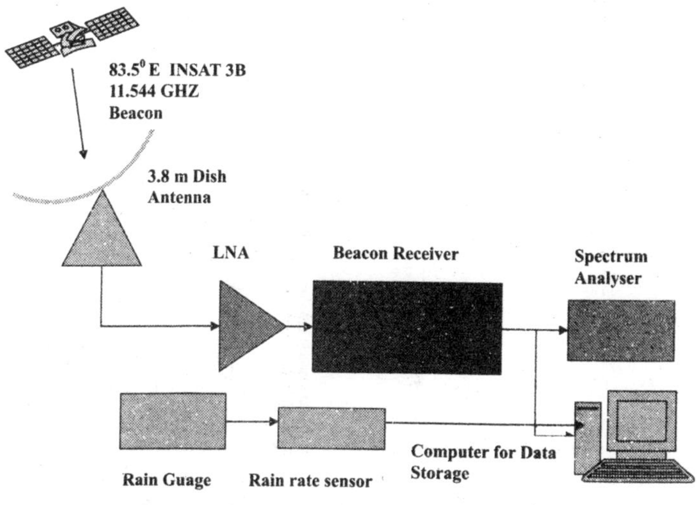

Figure 2: Experimental Set Up for INSAT 3B Signal Rain Attenuation Measurements

RESULTS AND DICUSSIONS

The prediction models *ITU-R (2003), Garcia- Lopez (1998). Moupfouma [1987(a)], and Dissanayake- Allunett-Haidra (DAH) (1997)* are used to estimate the effect of rain on 11.5 GHz signals. Cumulative distribution of rain rate and rain attenuation are estimated and compared with the above predicted models. However it should be mentioned here that the data used for the present study is only for three months and hence the studies are very preliminary. Examples of signal strength attenuation of INSAT- 3B Beacon signal at 11.54 GHz during high rain days are shown in Figures 3 to 8.

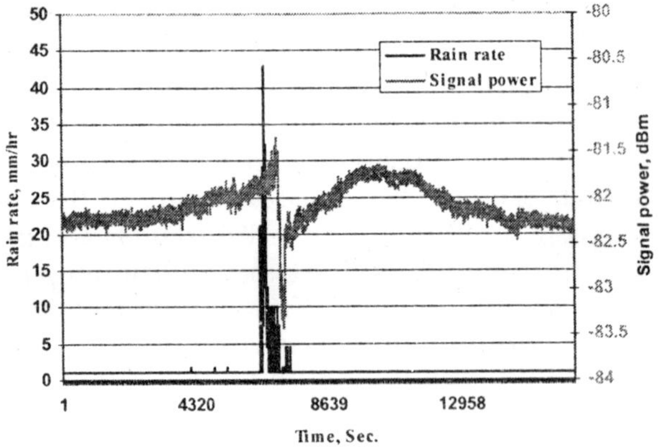

Figure 3: Rain Attenuation Observed on 6-9-2004

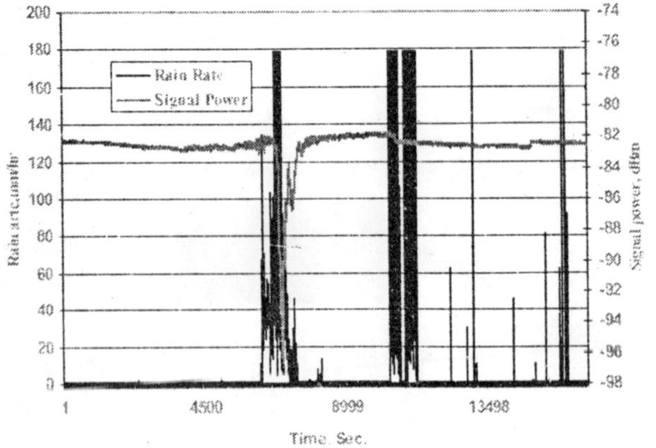

Figure 4: Rain Attenuation Observed on 13-9-2004

It can be observed from the figures that on 6-9-2004, the rain intensity was less. The maximum rain rate observed is about 45 mm/hr. Correspondingly the signal attenuation is very low and maximum attenuation is about 3 dB. The power spectral density of the signal power observed on 6-9-2004 is computed and is shown in figure 9. The power spectral density of a rain attenuation event is expected to decrease with frequency at about -20 dB^2 per decade for frequencies between about 10^{-4} Hz and 0.02 Hz *(Marticciani,1994;1996)*. At frequencies lower than 10^{-4} Hz, the spectral components are caused by long-term variations in the atmospheric conditions, and the corresponding spectrum is flat. Above about 0.02 Hz scintillations and fast fluctuations of rain attenuation are

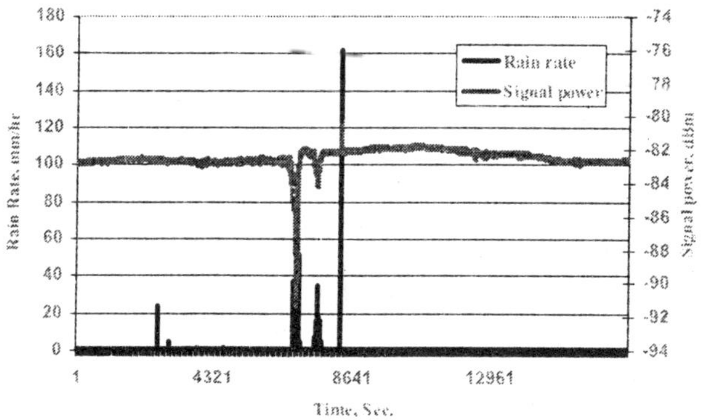

Figure 5: Rain Attenuation Observed on 25-9-2004

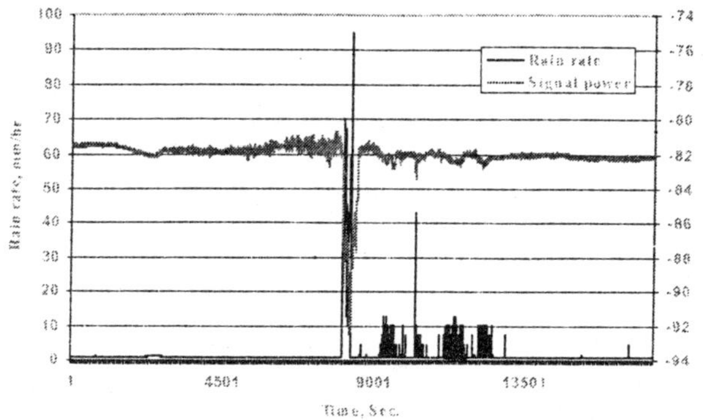

Figure 6: Rain Attenuation Observed on 1-10-2004

dominant. From the figure 9, it can be observed that between 10^{-4} Hz and 0.02 Hz the slope is about 20 dB^2 per decade where as rain induced scintillations are seen up to 0.08 Hz.

In the past, the separation of scintillations from attenuation has been carried out by high pass filtering. However, it is not possible to establish a unique filter cut off frequency for all weather conditions and system characteristics. This will be shown in the following discussions by analyzing the PSD (Power Spectral Density) plots of different rain events.

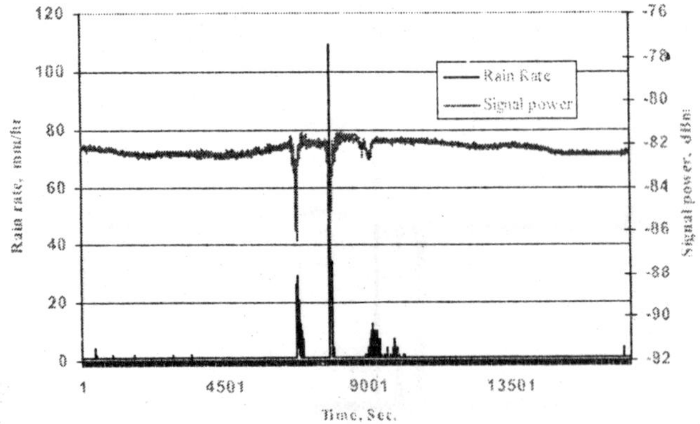

Figure 7: Rain Attenuation Observed on 2-10-2004

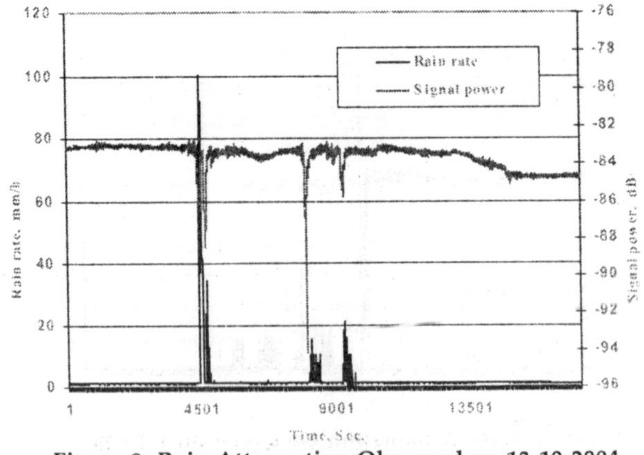

Figure 8: Rain Attenuation Observed on 13-10-2004.

Typical PSD plot observed on 13-9-2004 (figure 10) is quite different from the one observed on 6-9-2004. The rain rate observed during this event is quite high,180 mm/hr where in we can see clearly the rain attenuation slope upto about 0.01 Hz. Similarly one can find the presence of rain induced scintillations in PSD plots of 25-9-2004 & 1-10-2004 (Figurés 11 & 12). The power spectral density profile observed on 02-10-04 (Figure 13) is quite interesting as power fluctuations show some periodicity.

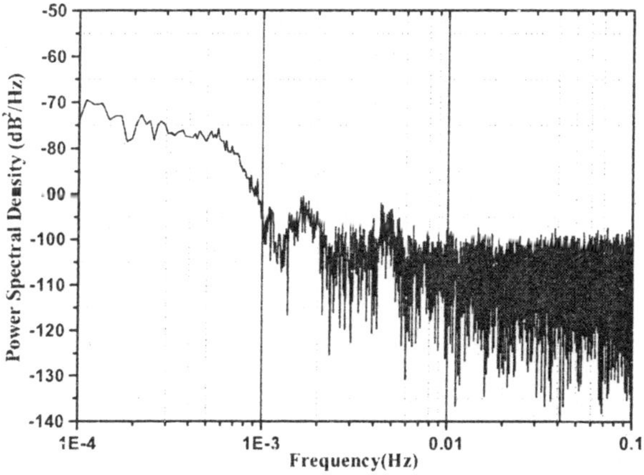

Figure 9: Power Spectral Density on 6-9-2004

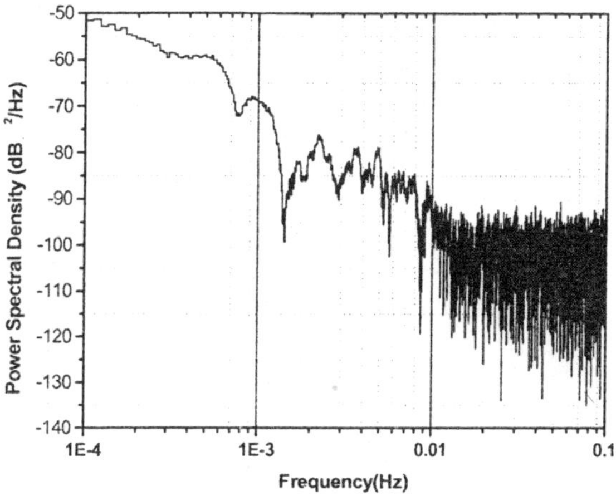

Figure 10: Power Spectral Density on 13-9-2004

In order to find out some correlation with the observed rain rate the FFT is performed for all the days for both signal power as well as rain rate and the results are shown in figures 14 and 15. From the figures one can observe a good periodicity in the rain events on 13-9-2004, 14-9-2004, 1-10-2004, 2-10-2004. On 6-9-2004 FFT of rain rate shows no periodicity.

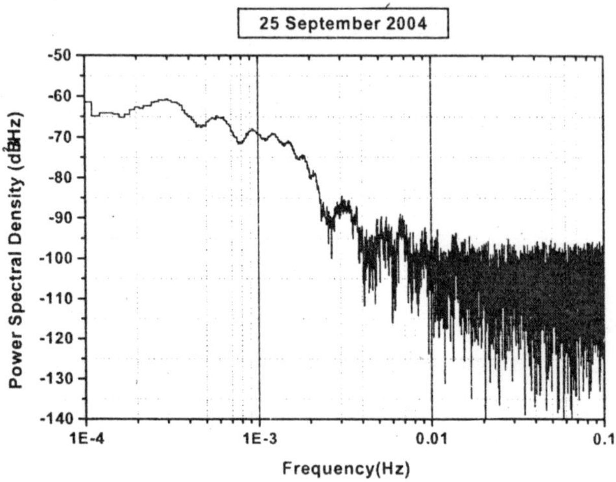

Figure 11: Power Spectral Density on 25-9-2004

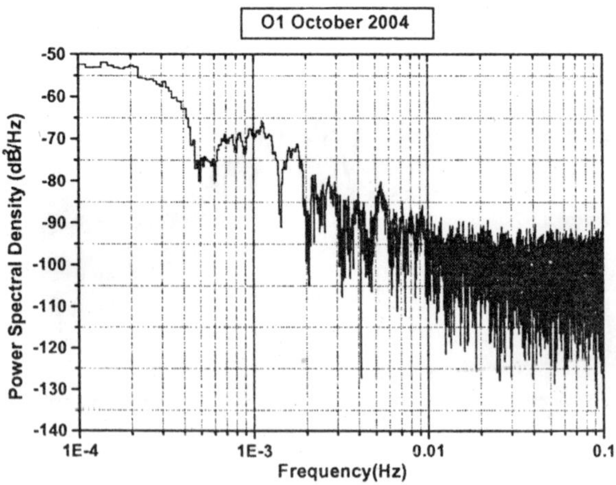

Figure 12: Power Spectral Density on 1-10-2004

It is well established that the rain attenuation is a function of frequency, elevation angle, polarization angle, rain intensity, rain drop size distribution and rain temperature. Fade

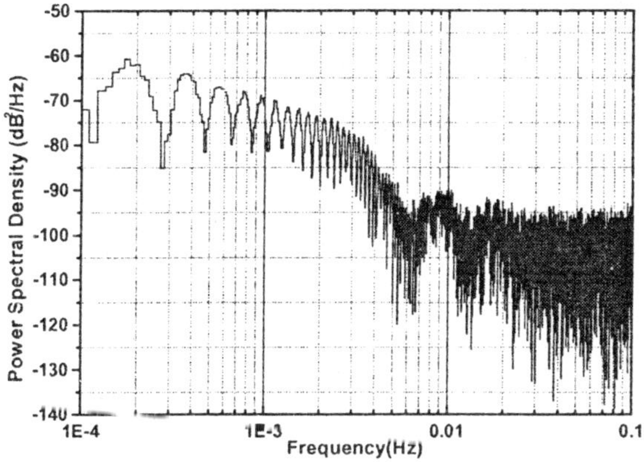

Figure 13: Power Spectral Density on 2-10-2004

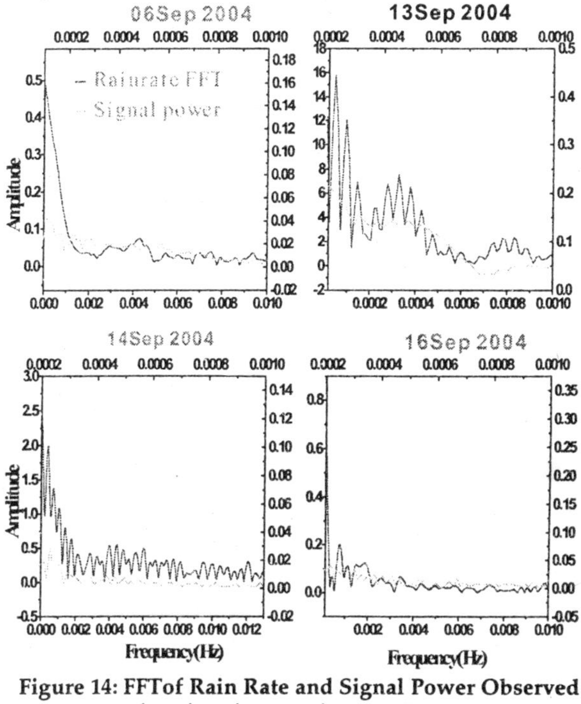

Figure 14: FFTof Rain Rate and Signal Power Observed on 6th, 13th, 14th and 16th September, 2004

durations and rates are closely correlated with rain type. Usually stratiform rains are associated with longer durations and slower fade rates. The stratiform rain is defined with rain rates below 6 mm/hr. Figure 16 shows the occurrence probability of rain rate in mm/hr during different months. It can be clearly seen that the probability of convective rain is more in September and October where as it is minimum, in August. Obviously this shows the observed fading is higher in the months of September and October. The same can be observed from the CDF graphs shown in Figure 17.

It is quite evident from the measurements made in extra tropical countries that the effect of cloud attenuation and scintillations are negligible at high elevation angles. However, our results show that the contribution of cloud attenuation and scintillations are significant at our latitudes even at higher elevation angles. The results given in Tables 4 to 8 in

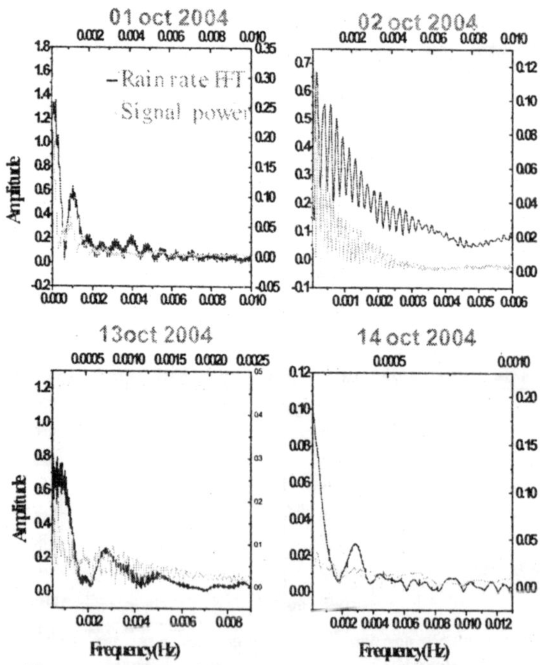

Figure 15: FFT of Rain Rate and Signal Power Observed on 1st, 2nd, 13th and 14th October 2004

Chapter-6 are produced by running the on-line ITU-R model for the present system, showing the contributions of cloud attenuation and scintillations at different percentages of probabilities.

Comparison with ITU-R Model

The objective of the present study is to evaluate the adoptability of ITU-R model to tropical countries like India. *ITU-R P.839-3 (2001)* has revised the rain maps and the same

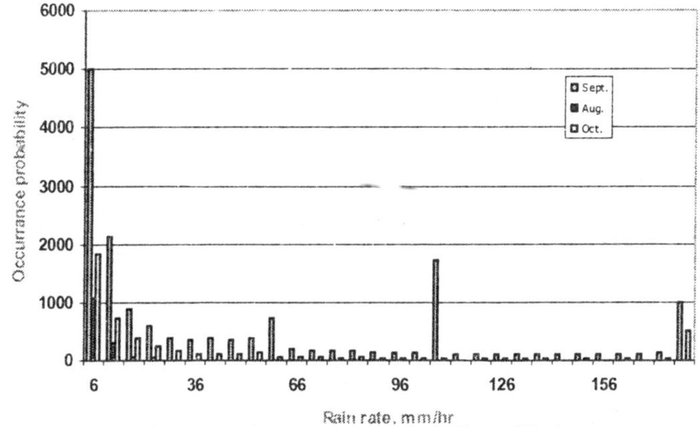

Figure 16: Occurrence Probability of Rain Rate for Aug. Sept. & Oct.2004.

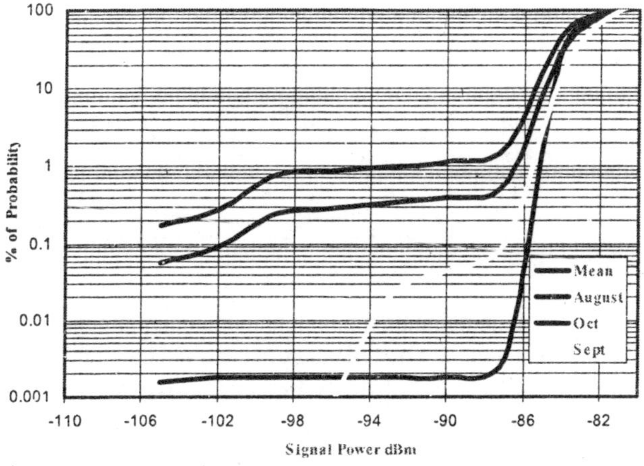

Figure 17: CDF for the Months Aug., Sept., & Oct. 2004.

is shown in figure 18. It can be observed that the present measurement location Hassan is having a rain rate of 80 to 100 mm/hr for 0.01% probability. The cumulative distribution of rain rate observed over Hassan over a period of one year is shown in figure 19. At 0.01% probability the rain rate is about 110 mm/hr which is close to ITU-R predictions for this location.

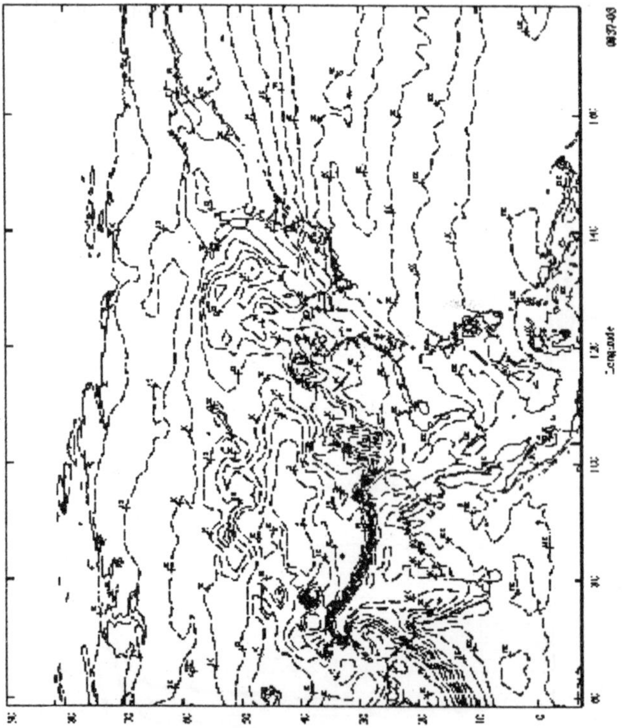

Rain attenuation calculated with different propagation models along with the observed results are shown in figure 20. A good correlation between different models is observed between 1 to 0.1% probability. Large deviation is observed between 0.01 to 0.001% probability levels. The observed attenuation is much higher than that predicted by the models. *ITU-R model* closely predicts the values compared to other models.

Adji et al., (1995) made studies on the rain attenuation characteristics at 11.2 GHz from Indonesian domestic satellite with 3.6 m diameter antenna and at 85^0 elevation. The reported attenuations exceeding 0.01% of time is around 15 dB and at 1% of time it is about 1 dB. In the present study, attenuations exceeding 0.01% of time is around 7 dB and at

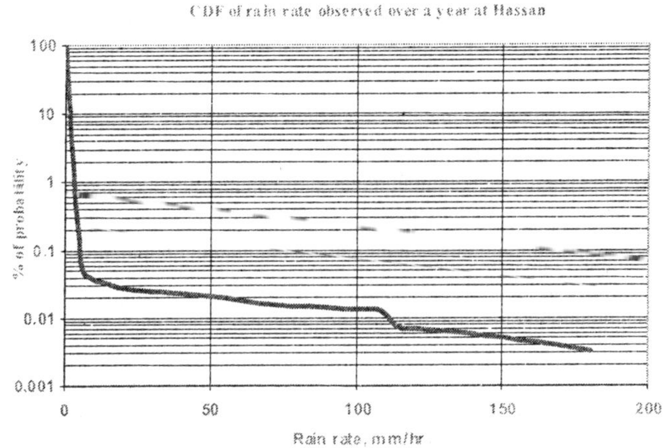

Figure 19: CDF of Rain Rate Observed Over a Year (2004) Over Hassan

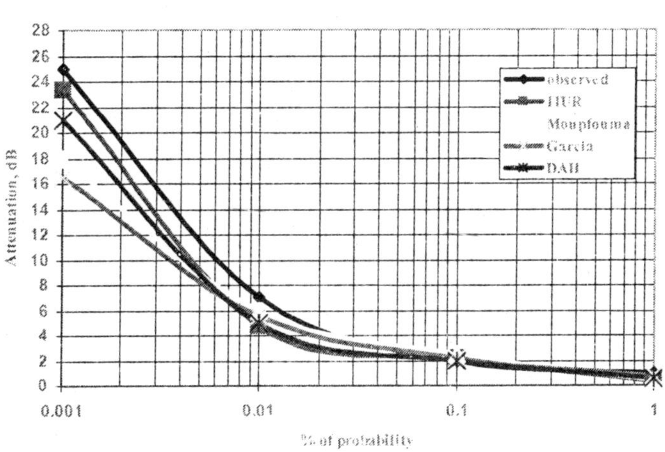

Figure 20: Rain Attenuation Predictions with Different Models

1% of time it is about 2 dB at 72^0 elevation angle. *Prasad et al., (1994)* made radiometrically measured rain attenuation over New Delhi at 10^0 elevation for 11 GHz, and reported 10 dB at 0.01% of time and around 5 dB for 0.1% of time. Measurements from other tropical countries *[McCathy et al., 1992(a), (b), (c)]* have appeared in Ku Band but a comparison of the present observations with those can not be carried out as their elevation angles are totally different and mainly are from radiometric measurements. The observed results from all the tropical countries deviates considerably with the ITU-R predictions. The remarkable point in all these results from the tropical countries is that the ITU-R prediction deviates from the observed attenuation values. These deviations are expected because the large drop diameters even at low rain rates will result in higher attenuation in tropical countries. The 0^0 C isotherm height or the effective rain height given by ITU-R is a fixed value. However the 0^0 C isotherm height value displays a seasonal and rain fall type dependence (*Ong.J.T et al., 2000)*. Detailed analysis of rain fall data, type of precipitation system and melting layer height over a long time period is required to arrive at statistically significant results.

CONCLUSIONS

Using the available rain rate and attenuation of signal power an attempt is made to study the rain attenuation characteristics over Hassan, a tropical Indian station using INSAT -3B satellite data. The observed rain fall and predicted attenuation are in agreement with ITU-R revised models. Earlier (*Rao et al., 1999)* comparisons with ITU-R deviated very much from the observed results. The observed attenuation is slightly higher than ITU-R estimated values.

CHAPTER–8

ATTENUATION OF KU & KA BAND SATELLITE COMMUNICATION SIGNALS DUE TO WATER VAPOUR ABSORPTION–PREDICTIONS USING RADIOSONDE DATA

INTRODUCTION

The increasing demands to use SHF and EHF bands in terrestrial radio systems and satellite communications especially in tropical and sub-tropical regions requires a prior knowledge of water vapour distribution and rain rate statistics for optimising the performance. For satellite communication links, it is very important to predict the atmospheric effects on Ku- and Ka-bands to ensure the designed levels of service availability in these frequency bands.

The atmosphere is a layer of gasses covering the surface of the earth. There are many gases in the atmosphere. The two that cause almost all of the radio wave attenuation are Oxygen (O_2) and water vapour (H_2O). While oxygen is an invariant parameter that can be easily accounted for and does not change with latitude or season, water vapour is highly variant; the diurnal and seasonal variability of water vapour can be extremely large. The electric dipole associated with the water vapour molecule interacts with the incident electromagnetic radiation to produce rotation absorption lines at 22.2, 183.3 and 324 GHz. The two components due to

Oxygen (O_2) and water vapour (H_2O) should be added together for the total specific attenuation. Hence, the gaseous absorption is calculated as the sum of water vapour absorption and oxygen absorption.

The water vapour absorption at Ku- and Ka-bands is considered for study in this chapter. Oxygen absorption is not considered, as the FSS (Fixed Satellite Services) do not employ the frequencies close to the oxygen absorption band (~60 GHz) at present in India. The attenuation of Satellite signals due to water vapour absorption is calculated using the formulas referred in *Recommendations ITU-R, P.676-5* and compared with the water vapour attenuation calculated using Radiosonde data collected from the India Meteorological Department (IMD) observatory at Hyderabad. Radiosonde data is collected for three years period starting from Jan. 2002 to Oct. 2004. Predictions of specific attenuation and path attenuation due to water vapour absorption are presented in this chapter.

Using the Radiosonde data, monthly and yearly variation of water vapour absorption for various frequencies and elevation angles are calculated and compared with the *Recommendations of ITU-R P.676-5* to analyse the attenuation pattern for Satellite communications in Ku and Ka Bands due to water vapour absorption. This study will provide better understanding of propagation issues by water vapour absorption loss in Ku-band and Ka-band satellite links operating in a tropical country like India, where in, not much work is reported in this area.

ATTENUATION DUE TO WATER VAPOUR AND RAIN

Molecular absorption between several GHz and several hundred GHz is primarily caused by oxygen and water vapour. The oxygen molecule has a small magnetic moment and absorption is produced by magnetic interaction with the incident field. Near 60 GHz and 119 GHz, many oxygen absorption lines merge together, at sea level pressures, to form a single broad absorption band. The molecular oxygen absorption at these two resonance frequencies is higher. Gaseous losses increase linearly in dB with distance. The loss due to water vapour varies with the absolute humidity level (*Howard Sandle, 1999*). The absorption loss due to water vapour near 22.235 GHz, 183.3 GHz and 325.4 GHz (resonance frequencies of water vapour) are maximum. Satellite communication links operating in the Ku-band (11/14 GHz) and the Ka-band (20/30 GHz) will suffer from absorption losses due to the water vapour and oxygen in the atmosphere. The interaction produces a family of rotation absorption lines in the vicinity of 60 GHz and an isolated line at 119 GHz (*Crane,1981*). The thermal emission at frequencies near the peaks of the water vapour lines is also proportional to the number of water vapour molecules and may be used to sense the water vapour concentration when the temperature is known.

Hydrometeors which include all products of condensed water vapour such as rain, hail, fog, ice, cloud and snow not only attenuate radio waves above 5 GHz but, also cause impairment to communication systems in a number of ways. Hail, fog, ice and snow are less significant in a tropical country like India, where as rain is the most serious form of hydrometeors which because of its large dielectric constant produces heavy displacement currents and hence absorption. The water droplet is essentially an imperfect conductor with a dielectric constant 80 times that of air. Also the displacement currents induced in the droplets of rain or fog are sources of scatter radiation and thus, in the direction of interest such scatter effectively causes 'attenuation'.

The most important effects of hydrometeors on radio wave propagation are (Ippolito, 1981)

(a) Attenuation caused by dissipation of radio wave energy as heat.

(b) Scatter resulting in loss in the desired direction and consequently causing interference to other systems. Scatter loss can dominate when the wavelength is much smaller than hydrometeor size.

(c) Depolarization due to the non-spherical nature of rain drops.

(d) Rapid amplitude and phase scintillations caused by equivalent Multi path propagation..

(e) Antenna gain degradation due to phase dispersion of ray paths reaching the antenna.

(f) Bandwidth coherence reduction especially in digital systems involving carriers spanning over large channel bandwidths.

This chapter is confined only to the water vapour absorption at Ku and Ka band frequencies.

SPECIFIC ATTENUATION DUE TO DRY AIR AND WATER VAPOUR

The specific attenuation due to dry air and water vapour can be estimated using the following simplified algorithms given in *Recommendation ITU-R, P.676-5* which are based on curve fitting to the line-by-line calculation. The input parameters required for the calculation include frequency, path elevation angle, and height above mean sea level and water vapour density. The oxygen attenuation is considered as background effect with very little temporal variation; variations in gaseous absorption arise from changes in the amount of water vapour in the atmosphere. In order to estimate the distribution on an annual basis, the annual distribution of water vapour density is required (*Asoka Dissanayake, 2002*). In *recommendation ITU-R, P.676-5,* the absorption due to oxygen and water vapour at the zenith, using the mean annual global reference atmosphere (Pressure of 1013 hpa, a temperature of 15^0 C and water vapour density of 7.5 g/m^3), has been predicted for Ka and Ku band and it is shown in the following table 1.

However, these values differ significantly in tropical countries where the temperature and humidity play a major role. Hence the objective of the study is to verify the adoptability of these recommendations to tropical countries, specially to India.

METHODOLOGY

Using the Recommendation ITU-R,P.676-5,(2001) together with the Recommendation ITU-R, P.453-8, the water vapour absorption is estimated using the methodology given below:

For water vapour, the specific attenuation, γ_w (dB/km) is given by,

$$\gamma_w = \left\{ \begin{array}{l} 3.13\times10^{-2} r_p r_t^2 + 1.76\times10^{-3}\rho r_t^{8.5} + \\ r_t^{2.5}\left[\begin{array}{l} \dfrac{3.84\xi_{w1} g_{22}\exp(2.23(1-r_t))}{(f-22.235)^2+9.42\xi_{w1}^2} + \dfrac{10.48\xi_{w2}\exp(0.7(1-r_t))}{(f-183.31)^2+9.48\xi_{w2}^2} \\ + \dfrac{0.078\xi_{w3}\exp(6.4385(1-r_t))}{(f-321.226)^2+6.29\xi_{w3}^2} + \dfrac{3.76\xi_{w4}\exp(1.6(1-r_t))}{(f-325.153)^2+9.22\xi_{w4}^2} \\ + \dfrac{26.36\xi_{w5}\exp(1.09(1-r_t))}{(f-380)^2} + \dfrac{17.87\xi_{w5}\exp(1.46(1-r_t))}{(f-448)^2} \\ + \dfrac{883.7\xi_{w5} g_{577}\exp(0.17(1-r_t))}{(f-557)^2} + \dfrac{302.6\xi_{w5} g_{752}\exp(0.41(1-r_t))}{(f-752)^2} \end{array} \right] \end{array} \right\}$$

$for f \leq 350 GHz$

$$\xi_{w1} = 0.9544 r_p r_t^{0.69} + 0.0061\rho$$

$$\xi_{w2} = 0.9544 r_p r_t^{0.64} + 0.0067\rho$$

$$\xi_{w3} = 0.9561 r_p r_t^{0.67} + 0.0059\rho$$

$$\xi_{w4} = 0.9543 r_p r_t^{0.68} + 0.0061\rho$$

$$\xi_{w5} = 0.955 r_p r_t^{0.68} + 0.006\rho$$

$$g_{22} = 1 + (f-22.235)^2/(f+22.235)^2$$

$$g_{577} = 1 + (f-557)^2/(f+557)^2$$

$$g_{752} = 1 + (f-752)^2/(f+752)^2$$

Where,

f = frequency (GHz)

r_p = p/1013 —(8.10)

r_t = 288/(273+t) —(8.11)

p : pressure (hPa)

t : temperature (°C) and

r : water vapour density (g/m^3).

Using the *Recommendation ITU-R,P.453-9 (2003)*, we can find the water vapour density "ρ". The relationship between water vapour pressure "e" and relative humidity "H" is given by,

$$e = H e_s / 100 \quad \text{hPa} \qquad \text{—(8.12)}$$

$$\text{and } e_s = a \exp(bt/(t+c)) \quad \text{hPa} \qquad \text{—(8.13)}$$

where,

H : Relative humidity (%)

t : Celsius temperature (°C)

e_s : saturation vapour pressure (hPa) at the temperature t (°C) and the co-efficient a, b, c are:

For water:

a = 6.1121

b = 17.502

c = 240.97

(valid between -20° to +50°,with an accuracy of ±0.20%)

For ice:

a = 6.1115

b = 22.452

c = 272.55

(valid between -50° and 0°,with an accuracy of ±0.20%)

The saturation vapour pressure, "e_s"(hPa) at the temperature t (°C), the co-efficient a, b, c for water is considered. Because the co-efficient a, b, c for water is valid between -20° to +50°,with an accuracy of ±0.20%.

Vapour pressure "e" is obtained from the water vapour density "ρ" using the following equation,

$$e = \rho T / 216.7 \quad \text{hPa} \qquad \text{—(8.14)}$$

By re-arranging the above equation, water vapour density "ρ" in g/m^3 can be found,

$$\rho = e^{*}216.7/T \text{ in } g/m^3 \qquad \text{—(8.15)}$$

Table 1
ITU-R P.676-5 Results for Zenith Oxygen and Water Vapour (7.5 g/m³) Absorption

Frequency (GHz)	*Oxygen Absorption at Zenith (dB)*	*Water Vapour Absorption at Zenith (dB)*
10	0.043	0.013
20	0.056	0.25
30	0.1	0.17
40	0.25	0.17

SLANT PATH ATTENUATION

The gaseous attenuation model given in *Recommendation ITU-RP.676-5* , can be used to predict the water vapour absorption for terrestrial and slant paths. It defines a simple algorithm for the calculation of water vapour absorption along the slant paths through the earth's atmosphere in the frequency range 1-350 GHz.The zenith path water vapour absorption is obtained by multiplying an equivalent height for water vapour "h_w" with the specific attenuation due to water vapour, γ_w (dB/km) calculated in the equation (8.1).

For water vapour, the equivalent height is,

$$h_w = 1.65\left\{\begin{matrix} 1+\dfrac{1.61}{(f-22.23)^2+2.91}+\dfrac{3.33}{(f-183.3)^2+4.58} \\ +\dfrac{1.90}{(f-325.1)^2+3.34} \end{matrix}\right\} km$$

in km for f ≤ 350 GHz. —(8.16)

The total zenith attenuation is,

$$A = \gamma_w h_w \text{ (dB)} \qquad \text{—(8.17)}$$

For an elevation angle "j" between 5° and 90°, the slant path attenuation (for earth-space (paths) is obtained using the following relation

$$A = \gamma_w h_w / \sin \varphi \text{ (dB)} \qquad \text{—(8.18)}$$

Where, "φ" is the elevation angle.

Path attenuation for satellite communication links with any geometrical configuration due to water vapour absorption can be predicted by dividing the atmosphere into horizontal layers, specifying the profile of the meteorological parameters such as pressure, temperature and relative humidity along the path. Using these meteorological parameters with the procedures given in *Recommendation ITU-R, P.676-5* and *Recommendation ITU-R, P.453-9* , the specific attenuation and hence the path attenuation due to the water vapour can be predicted.

The given path is divided in to horizontal layers of 300 m thickness in altitude, with flat earth assumption and uniform layers. The profile of meteorological parameters (pressure, temperature and relative humidity along the path) of each layer is specified by the radiosonde data of Indian Meteorological Department. The water vapour density "ρ" in gm/m^3 for each layer is calculated by using *Recommendation ITU-R, P.453-9*. The specific attenuation due to water vapour absorption "γ_w"(dB/km) and path attenuation for each layer by using the procedures given in *Recommendation ITU-R, P.676-5* are calculated. The summation of each layer's path attenuation will give the "total path attenuation due to water vapour absorption". Similarly we can predict the path attenuation due to water vapour absorption for different frequencies and elevation angles.

WATER VAPOUR ABSORPTION ANALYSES

Monthly and Yearly Variability for Water Vapour Absorption

For the design of terrestrial microwave line-of-sight systems and satellite system and High Altitude Platform Systems (HAPS) operating in the higher microwave bands, the water vapour absorption loss along the path must be taken in to account. For system planning there is a need to quantify the water vapour absorption loss over any month or over a year.

In this chapter analyses are made using radiosonde data collected from India Meteorological Department over a period of three years (2002,2003 and 2004). The morning radiosonde ascent starts at 0000 GMT and the evening radiosonde ascent at 1200 GMT. The maximum altitude considered in this study is 5.0 km as the content of water vapour present above this altitude can be neglected for our purpose.

In this chapter analyses are made for the monthly and yearly variations of water vapour absorption for various frequencies and elevation angles. The mean zenith water vapour absorption at frequency f=15 GHz and at f=30 GHz is computed and compared with the Recommendation ITU-R, P.676-5.

Monthly Variations

The zenith water vapour absorption at frequency f=15 GHz is computed for three years (i.e., for the years 2002, 2003 & 2004) using the radiosonde data provided by the India Meteorological Department. The monthly Cumulative Distribution Function (CDF) curves for the morning and the evening radiosonde data are plotted and analyzed. However it is found that the difference in the zenith water vapour absorption between the morning and evening derived from radiosonde data is negligible and hence the morning and evening radiosonde data are combined together and processed.

Combined CDF Graph of Morning and Evening Radiosonde Data

The water vapour absorption separately for each of the twelve months for three years (2002 to 2004) using the morning and the evening radiosonde data combined together and the CDF curves plotted are shown in the figure 1. The water vapour absorption is found to be between 0.37 dB and 0.97 dB. For 50 percent of time, the water vapour absorption was greater than 0.68 dB for July and for December it was greater than 0.78 dB. For 10 percent of time, the water vapour absorption was greater than 0.79 dB for March and for June, it was greater than 0.82 dB. Similarly for 1 percentage of time, the water vapour absorption was greater than 0.85 dB for January and for September, it was greater than 0.95 dB.

Yearly Variations

Since there is no significant difference of water vapour absorption, three years of radiosonde data are combined together to study the yearly variation of water vapour absorption. The following Figure 2 shows the CDF curves for

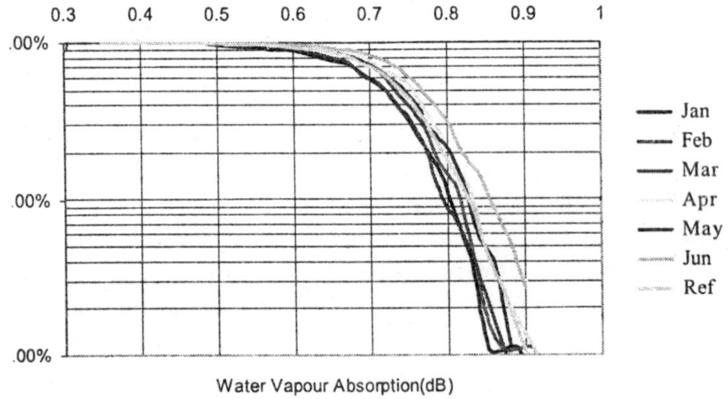

Figure 1 (a): Monthly Variation of Water Vapour Absorption Using the Morning and the Evening Data Jan. to Jun. (2002 to 2004)

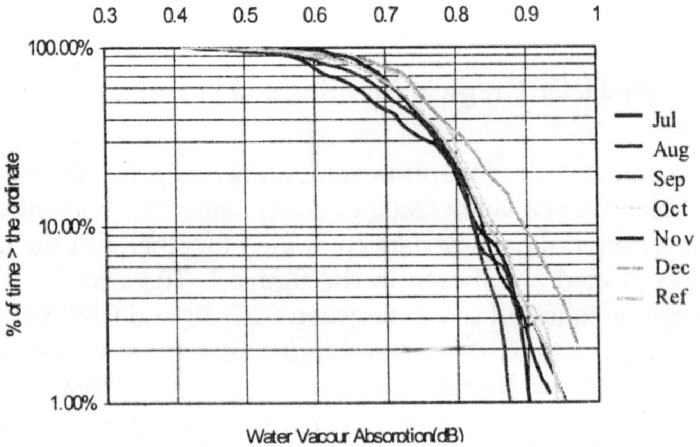

Figure 1 (b): Monthly Variation of Water Vapour Absorption using the Morning and the Evening Data Jul. to Dec. (2002 to 2004)

three years (2002, 2003 and 2004). It can be seen from Figure 2, that the year 2004 has more water vapour absorption than other two years. It can be seen from Figure 2, that the water vapour absorption for 2002 and 2003 was almost same for 85% of time and for the remaining time the water vapour absorption for 2002 was less than for 2003. But the difference in water vapour absorption between 2002 and 2003 is very minimum. For 50 percentage of time, the water vapour absorption is greater than 0.72 dB for 2002 and 2003 and it is greater than 0.77 dB for 2004, for 10 percentage of time, the

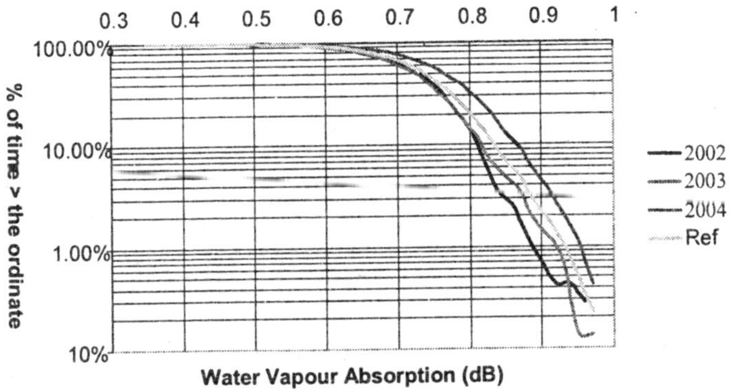

Figure 8.2.Yearly Variation of Water Vapour Absorption using the Morning and the Evening data (2002 to 2004).

water vapour absorption is greater than 0.81dB for 2002 and it is greater than 0.82 dB for 2003, and it is greater than 0.86dB for 2004. For 5 percentage of time, the water vapour absorption is greater than 0.82 dB for 2002 and it is greater than 0.85 dB for 2003 and it is greater than 0.9 dB for 2004. For 1 percentage of time, the water vapour absorption is greater than 0.89 dB for 2002 and it was greater than 0.92 dB for 2003, and it is greater than 0.95 dB for 2004. For 0.5 percentage of time, the water vapour absorption is greater than 0.91dB for 2002 and it is greater than 0.93 dB for 2003, and it is greater than 0.96 dB for 2004. It is found that the highest yearly variation of water vapour absorption among these three years was 0.08 dB, which is less than the highest monthly variation of water vapour absorption (0.1dB).

COMPARISON OF WATER VAPOUR ABSORPTION WITH ITU-R RESULTS

The yearly mean zenith water vapour absorption using the three years of radiosonde data computed at frequencies f=15 GHz and at f=30 GHz is 0.07 dB and 0.17 dB respectively. The *recommendation* :ITU-R, P.676-5 has predicted the zenith water vapour absorption using the mean annual global reference atmosphere that is given in the recommendation :ITU-R, P.835 , at frequency, f=15 GHz and at 30 GHz is 0.25 dB and 0.17 dB respectively. The comparison between the ITU-R results and our predictions are shown in the Table 2. It can be seen from these results that the observed water vapour absorption is more than the *ITU-R, P.675-5* predictions. The difference between our predictions and the *ITU-R, P.676-5* predictions are due to the difference in the surface level water vapour density used. The surface level water vapour density used in the *ITU-R, P.675* is 7.5g/m^3 whereas the surface level water vapour density for Hyderabad is 20 g/m^3.

Table 2
Comparison of Water Vapour Absorption Between ITU-R Results and Our Predictions

Frequency (GHz)	*ITU-R Results Zenith Water Vapour Absorption (dB)*	*Our PredictionsMean Zenith Water Vapour Absorption (dB)*
15	0.07	0.74
30	0.17	0.72

WATER VAPOUR ABSORPTION FOR VARIOUS ELEVATION ANGLES

It is required to know the variation of water vapour absorption for various elevation angles because the satellite links are operating at different elevation angles depending on the satellite location and the earth station site. This section analyses the water vapour absorption (at 15 GHz) at elevation angles, 5°, 10°, 15°, 30°, 45° and 90° for June 04. June month is selected as we have found that it is the worst month among all the three years in the above analysis. It is also found that the year 2004 is the worst year among the three years considered in this study. The following Figure 3 shows the water vapour absorption at elevation angles, 5°, 10°, 15°, 30°, 45° and 90° for June 04 .It can be seen from the same figure that the water vapour absorption at the elevation angle, 5° is much higher as expected than other elevation angles. It is found that the mean water vapour absorption at 5° is 9.79 dB, at 10° is 4.92 dB, at 15° is 3.3dB, at 30° is 1.71dB at 45° is 1.21dB and at 90°(zenith) is 0.85 dB. Also the maximum and minimum water vapour absorption at the elevation angles, 5°, 10°, 15°, 30°, 45° and 90° for June 04 is found and listed in the following Table 3.

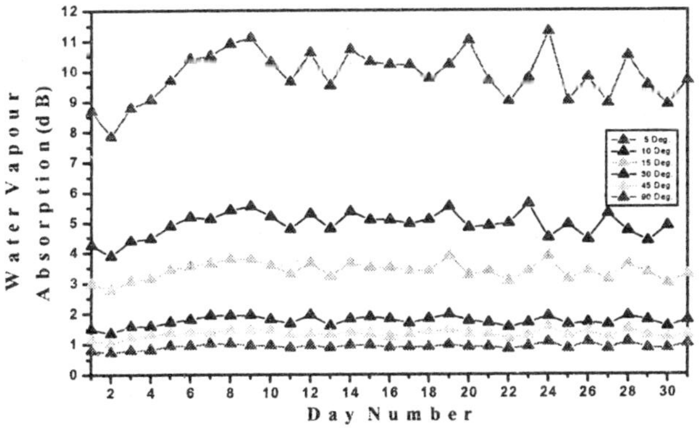

Figure 3: Water Vapour Absorption for Various Elevation Angles-June, M04.

The water vapour absorption increases as the elevation angle departs from the zenith. Because at the lower elevation angles, the higher portion of the path travelled by the signal propagates through the lower layers, which has higher water vapour concentration. That's why the water vapour absorption is much higher at the elevation angle 5° than other elevation angles. We can conclude that the mean water vapour absorption (at 15 GHz) for a worst month at the elevation angle 5° is found as 9.79 dB, at 10°, it is 4.92 dB, at 15°, it is 3.3 dB, at 30°, it is 1.71 dB, at 45°, it is 1.21 dB and at 90°, it is 0.85 dB.

Table 3
Water Vapour Absorption Observed in June 2004 for Different Elevation Angles

Elevation Angle	*Mean Water Vapour Absorption (dB)*	*Maximum Water Vapour Absorption (dB)*	*Minimum Water Vapour Absorption (dB)*
5°	9.79	11.22	7.86
10°	4.92	5.63	3.94
15°	3.3	3.78	2.65
30°	1.71	1.96	1.37
45°	1.21	1.38	0.97
90°(Zenith)	0.85	0.98	0.68

WATER VAPOUR ABSORPTION FOR VARIOUS FREQUENCIES

The atmospheric propagation characteristics of radio signals at a given frequency must be understood before designing communication links for that frequency. The water vapour absorption in Ku-band, Ka-band and oxygen band is considered in this analysis. The zenith water vapour absorption for different frequencies in Ku-band (for f=11 GHz & 14 GHz), Ka-band (for f=20 GHz, 22.235 GHz, 30 GHz), HAPS frequency (47 GHz) and oxygen band (f=60 GHz) is computed for the year 2004. The zenith water vapour absorption at sea level has been predicted in *Recommendation ITU-R, P.676-5* and it is shown in Figure 4. for a humidity of 7.5 g/m^3 The water vapour absorption varies appreciably with water vapour content. It can be seen from the figure that the water vapour absorption is higher at 22.235 GHz, because it is a resonance frequency of water vapour absorption line.

Figure 5 shows the CDF curves of zenith water vapour absorption for various frequencies in Ku-band, Ka-band and oxygen band. It can be seen from the figure 5, that the water vapour absorption at f=22.235 is higher than other frequencies. It is found that the zenith water vapour absorption for the frequency f=22.35 GHz was between 1.88 dB and 4.44 dB. For 50 percentage of the time the water vapour absorption was greater than 3.5 dB, for 10 percentage of the time the water vapour absorption was greater than 3.9 dB, for 5 percentage of time, the water vapour absorption was greater than 4.1 dB and for 1 percentage of the time it was greater than 4.26 dB. It is found that the water vapour absorption at frequency f=20 GHZ was between 0.79dB and 1.64 dB.For 50 percentage of the time the water vapour absorption was greater than 1.3 dB, for 10 percentage of the time the water vapour absorption was greater than 1.45 dB, for the 5

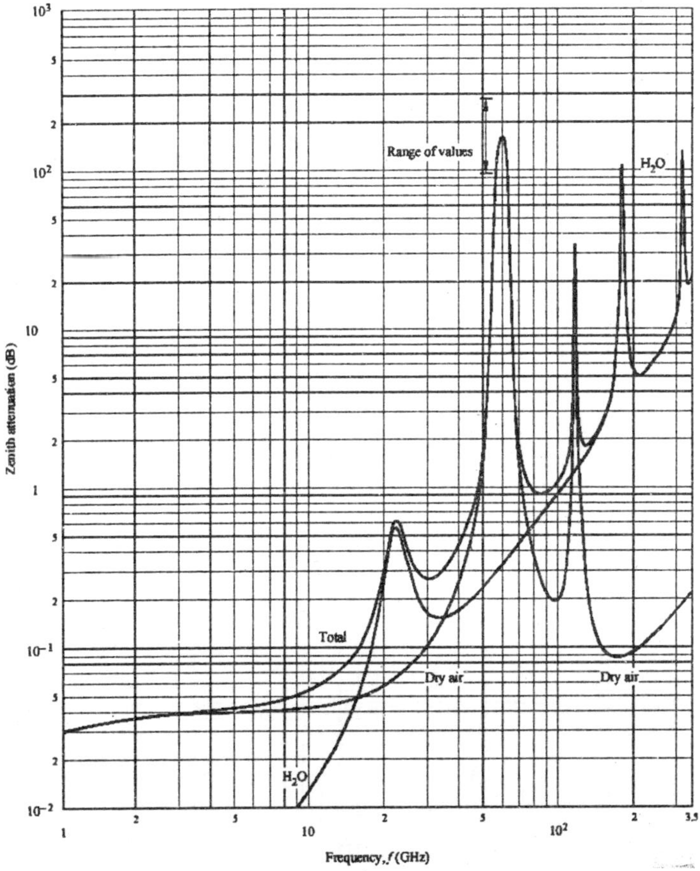

Surface pressure: 1 013 hPa
Surface temperature: 15° C
Surface humidity: 7.5 g/m³

Figure 4: Rec. ITU-R P.676-5 Water Vapour Absorption at Zenith from Sea Level

percentage of time the water vapour absorption was greater than 1.5 dB and 1 percentage of the time it was greater than 1.6 dB.

It is found that the zenith water vapour absorption for the frequency f=30 GHz was between 0.350 dB and 0.987 dB. For 50 percentage of the time the water vapour absorption was greater than 0.78 dB, for 10 percentage of the time the water vapour absorption was greater than 0.83 dB, for 5 percentage of time it was greater than 0.89 dB and for 1 percentage of time it was greater than 0.9 dB. It is found that the zenith water vapour absorption for the frequency f=33 GHz was between 0.4 dB and 0.93 dB. For 50 percentage of the time the water vapour absorption was greater than 0.72 dB, for 10 percentage of the time the water vapour absorption was greater than 0.8 dB, for 5 percentage of time it was greater than 0.82 dB and for 1 percentage of time it was greater than 0.9 dB.

It is found that the zenith water vapour absorption for the frequency f= 47 GHz was between 0.57 dB and 1.77 dB. WRC (World Radio Conference)-1997 has already designated the frequency around 47 GHz for the fixed services (FS) of high-altitude platform stations *[HAPS Reference]*. For 50 percentage of the time the water vapour absorption was greater than 1.04dB, for 10 percentage of the time the water

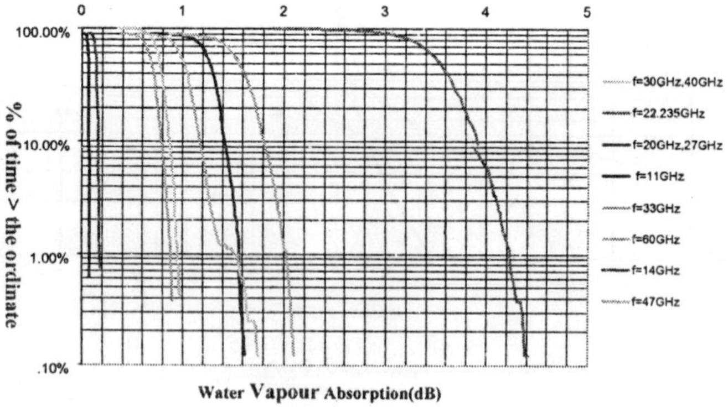

Figure 5: The CDF Curves of Water Vapour Absorption for Various Frequencies for the Year 2004.

vapour absorption was greater than 1.2 dB, for 5 percentage of time it was greater than 1.26 dB and for 1 percentage of time it was greater than 1.58 dB.

It is also found that the zenith water vapour absorption for the frequency f=60 GHz (oxygen band) was between 0.88 dB and 2.13 dB. For 50 percentage of the time the water vapour absorption was greater than 1.6 dB, for 10 percentage of the time the water vapour absorption was greater than 1.8 dB, for 5 percentage of time it was greater than 1.9 dB and

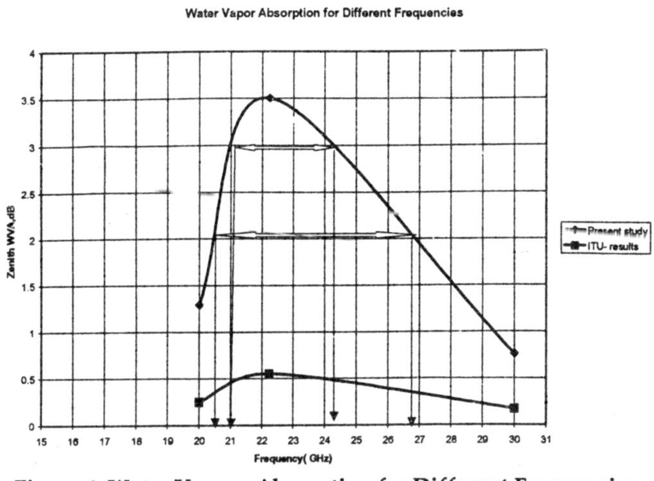

Figure 6: Water Vapour Absorption for Different Frequencies

for 1 percentage of time it was greater than 2 dB. It is found that the zenith water vapour absorption at the frequency f=14 GHz was between 0.09 dB and 0.21 dB. For 50 percentage of the time the water vapour absorption was greater than 0.16 dB,for the 10 percentage of time; the water vapour absorption was greater than 0.2 dB and for 5 percentage of the time it was greater than 0.2 dB. It is found that the water vapour absorption at the frequency, f=11 GHz was between 0.04 dB and 0.1 dB. For the 50 percent of time; the water vapour absorption was greater than 0.08 dB.

The yearly mean zenith water vapour absorption for 2004 is computed at the frequencies, f = 11 GHz, 14 GHz, 20/27GHz, 22.235 GHz, 30 GHz, 33 GHz, 47 GHz and 60 GHz and it is shown in the table 4. It is found that the mean zenith water vapour absorption at frequency, f=22.235 GHz is highest at 3.5 dB and at frequency, f=11 GHz it is 0.08 dB.

Table 4
Yearly Mean Zenith Water Vapour Absorption for Various Frequencies for 2004

Frequency (GHz)	*Mean Zenith Water Vapour Absorption (dB)*
11	0.08
14	0.16
20	1.3
22.235	3.5
30	0.76
33	0.71
47	1.04
60	1.6

Figure 6 shows the zenith water vapour absorption at different frequencies in Ku-band. The ITU-R curve also included. This curve can be used for approximate prediction of water vapour absorption in India. It can be seen from the figure 6 that the zenith water vapour absorption at 22.235 GHz is 3.5 dB and between 21GHz and 24.2 GHz it is 3 dB and between 20.5 GHz and 26.8 GHz the zenith water vapour absorption is 2 dB. It is found that the zenith water vapour absorption at 20 GHz is 1.29 dB and at 30 GHz is 0.76 dB.

SUMMARY OF THE CHAPTER

The monthly variations and the yearly variations of water vapour absorption are analyzed by using the three years of radiosonde data (2002 to 2004). The monthly variations of water vapour absorption and the CDF curves are studied. It is found that the highest monthly variation of water vapour absorption for the morning data and the evening data is 0.1dB. Since the difference between the morning and the evening water vapour absorption is not very significant, the morning and the evening data are combined together and analyzed. The yearly variations of water vapour absorption for the three years (2002 to 2004) also analyzed. It is found that the year

2004 has more water vapour absorption than other two years and the year 2003 has the lowest water vapour absorption among these three years. But the difference in water vapour absorption between the year 2003 and 2004 is very minimum.

The highest monthly variation of mean water vapour absorption among these three years, is found to be 0.18dB. The yearly mean zenith water vapour absorption using the three years radiosonde data has been computed at frequency, f=15 GHz and at f=30 GHz and compared with the *Recommendation: ITU-R P.676-5* predictions. It is found that our predictions are higher than *the ITU-R P.676-5* predictions. However it should be mentioned here that ITU-R User global reference atmosphere of surface pressure 1013 hpa, surface temperature of 15^0 C and water vapor density of 7.5 g/m^3. These values significantly differ from the observed values (near surface temperature 30^0 C and water vapour density of 20 g/m^3).

The water vapour absorption (at 30 GHz) at the elevation angles, 5°, 10°, 15°, 30°, 45° and 90° is analyzed. The water vapour absorption increases as the elevation angle departs from the zenith. Because at the lower elevation angles, the higher portion of the path travelled by the signal propagates through the lower layers, which has higher water vapour concentration. It is found that the mean water vapour absorption (at 30 GHz) at the elevation angle 5° is 9.794 dB, at 10° it is 4.92 dB, at 15° it is 3.3 dB, at 30° it is 3.3 dB, at 45° it is 1.207 dB and at 90° it is 0.853 dB.

The water vapour absorption for various frequencies in Ku (11,14 GHz), Ka-band (20, 22, 30 and 33 GHz), HAPS frequency (47 GHz) and oxygen band (60 GHz) is analyzed for the year 2004 as it has more water vapour absorption than 2002 and 2003. These results are useful for planning of future satellite communication system in India.

CHAPTER–9

SUMMARY AND CONCLUSIONS

The growing demand for new services and for higher data rates in existing services, and the convergence of communications and computing networks are placing unprecedented demands on the use of radio frequencies. The current surge of new techniques and systems being applied to telecommunications requires new radio propagation information, especially for Satellite-to-Earth paths, short range propagation within buildings, long range propagation into buildings and vehicles, and the influence of climate and ionospheric parameters on propagation modeling in general. These and other issues are of interest to URSI (International Union of Radio Science) and of concern to the ITU-R. Studies carried out in tropical countries assume great significance as not much input is provided from these countries to ITU-R.

India has a large population scattered in small rural communities and HF band has a premium and very crowded. In urban environments VHF/UHF services have prolifitated in recent years. Satellite services are expanding and plans exist to go into the Ku band soon. The Indian Ionosphere is characterized by large spatial and temporal gradients while the troposphere is humid, hot and subject to heavy rain. The theory and practice of Radio communications, especially with digital systems dominating every mode of communications, it is essential to estimate precisely the problems caused by the propagating medium which will decide the reliability of the link.

The tropospheric environment is highly dynamic in tropical countries and is very often dominated by frequent occurrences of sub-refraction, super refraction and ducting due to high humidity and high temperatures. It is well known that at low temperatures the variation in refractive index is very little for whole range of humidity values from zero to saturation, in contrast to great variations in the refractive index values at higher temperatures for similar changes in humidity. From this respect one cannot adopt the results from high latitude countries, where significant work might have been made on radio wave propagation, to the tropical countries. Detailed measurements will alone provide information about the deviations in results from ITU-R as ITU-R models in general adopt the results from extra tropical countries.

In this thesis, studies on the atmospheric constraints in HF/UHF/Satellite Communication in the Indian sub-continent are presented. Studies include HF fading, Atmospheric radio noise measurements, Atmospheric scintillations, water vapor absorption and rain attenuation in relation to several communication networks including those operated by certain Government Organizations in India. The thesis is organized in the following:

The thesis consists of a total of nine chapters dealing with HF/UHF propagation, satellite communication, rain and water vapor absorption and scintillations due to tropospheric propagation anomalies. *Chapter I* is a general introduction to the thesis and explicitly presents the details of communication problems in the tropics at HF/UHF and micro waves. The chapter also deals with attenuation due to hydrometers in view of the present demand for frequency reuse systems. The chapter describes briefly the overall status of communication problems in the tropics.

Chapter II presents briefly the HF, VHF, UHF and Satellite communication networks deployed in Government organizations in India. The specific requirements of HF, VHF, UHF and Satellite communication networks in different areas are also discussed. Specifications and details of the communication networks in the present and future are also discussed. The chapter is intended to synthesize a mix of several types of services and optimization for a fail-safe combination for priority Government users.

Atmospheric Radio Noise (ARN) is called static noise and is caused by the naturally occurring inter cloud electrical discharges in thunderstorms accompanying electrical flashes in the earth's atmosphere. They are spread over a wide range of radio spectrum and their intensity bears an inverse ratio to the frequency. Their effects are significant up to 30 MHz and hence Atmospheric Radio Noise (ARN) is a crucial parameter in limiting the efficiency of Radio Systems operating in HF range and it is in this context planned measurements of radio noise were made in India at National Physical Laboratory (NPL) New Delhi and at S.V. University, Tirupati. Results of this systematic study of ARN levels carried at NPL and at Tirupati, vis-à-vis earlier related studies are presented in *chapter–III*. It is observed that the NPL and Tirupati measured values are generally higher on all the frequencies, when compared with CCIR predicted values from the report 322. NPL, AIR and Tirupati measured values do not exhibit the dramatic increase in noise levels during the transition period (1600-2000 hrs IST) in contrast to the CCIR predicted values. This study has shown that there are certain major disparities between the actual ARN observations and CCIR predictions for India. The diurnal variation in the atmospheric radio noise (2.5–10.0 MHz) over Indian sub-continent is in total disagreement with the variations obtained at higher latitudes. This rather puzzling phenomenon has been interpreted through the present results, to be due to path attenuation differentials between the LOS mode and the ionospheric mode from the lightning source because of significantly greater cloud height in the Indian region, thus increasing the LOS range from clouds.

Chapter IV presents an exhaustive study of the short term HF fading characteristics over several long distance paths including BBC, Radio Sri Lanka, Mumbai, Chennai etc., recorded at Tirupati, New Delhi and Poona. It is observed that signals from Radio Sri Lanka seem to be fading at a faster rate as compared to those from BBC monitored at Tirupati. From the nature of short-term fading observed at Tirupati (13.5N, 79.5E), lasting from a second to a few seconds, it appears to be predominantly due to multi-path interference. Earlier studies made in India also support this view. The present study on fading in HF signals indicates that fading allowances of 11-25 dB would be required for 90% reliability of HF radio signals. In an earlier study a fading allowance of 9 dB was considered adequate for 95% reliability for comparatively higher latitudes in India. Also, a pronounced increase in the speed of fading is observed for locations close to geomagnetic equator. A comparison of measured field strength in the southern region of India and ITU-R predicted values indicates that the discrepancies can be significant in majority of cases; the predicted values are higher than the observed ones. Discrepancies as large as 8 dB have been observed. The present study indicates that fading allowance of 11-12.5 dB would be required for 90% reliability of HF radio Signals in the 10-18 N latitude ranges. ITU-R models and formulae in general based on global data appear to be inadequate for the Indian region primarily because of scarce data inputs from this region. This study assumes greater importance in the present day scenario because of the introduction of new digital audio modulation techniques in the HF band. Details of signal monitoring and data analysis are presented in the chapter. The validity of results obtained from measurements made at NPL, New Delhi and similar results recorded at Pune were compared. The most probable value for additional loss works out to be 8 to 12 dB. These are in conformity with the earlier observations monitored by National Physical Laboratory, New Delhi.

In mobile communications one of the major parameters of interest is path loss. Prediction of path loss assumes greater significance in view of the constantly changing environment conditions. There is still a great need for more narrowband and wideband measurements at frequencies of interest to put proper bounds on the statistical values. Any prediction model which is useful for most applications must consider all the important modes of propagation; mode selection must consider the extent to which the method is convenient to use. Irrespective of the prediction models deployed, estimated values always have to be verified by measurements and if necessary correction factors have to be derived and introduced in further predictions. To identify suitable narrowband methods in Southern India and to study various propagation phenomena affecting mobile radio communications, propagation measurements were conducted around 800 MHz. Observed results were converted into field strength values and are compared with different path loss prediction methods given by Hata, Egli, ITU-R, and Lee and the results are discussed in *Chapter-V*. It is observed that in urban environments the Hata-urban and ITU-R models shows a mean error of 6-7 dB compared to other models and are consistent with the earlier results observed in similar environment. A realistic site-specific propagation models are required with accurate terrain and building height information. Nevertheless Hata model in general provides reasonably good estimates for mobile communication planning.

For Satellite communication services in future, there is a need to deploy the higher Ku-band (11-18 GHz) and even higher Ka-band frequencies (20-30GHz) for providing higher data transmission rates, reduced interference potential and smaller receiving stations and antennas, leading to greater mobility. Use of smaller earth terminals for transportable applications like VSATs (Very Small Aperture Terminal), make it difficult to provide sufficient link margin for propagation related outages. Hence, reliable prediction of propagation impairments for low-availability systems becomes important. One of the main considerations while using high frequency bands in Ku- and Ka-bands, is the influence of the atmosphere on radio wave propagation. Since the influence of the atmosphere on microwave propagation increases with the frequency, a detailed analysis has to be made, so that these higher bands can be used for high quality mobile Satellite services.

Propagation impairments affecting Ka-band satellite links include tropospheric scintillations, gaseous absorption, cloud attenuation and rain attenuation. In this thesis the effect of Tropospheric scintillation on K_u Band Satellite Communication Link of INSAT-3B at 11.7 GHz, monitored at INSAT Master Control Facility (MCF), Hassan (13.07N; 76.08E) was studied and the results are presented in *chapter-VI*. Scintillations are produced due to the refractive index variations. A relative measure of these values can be obtained from structure intensity parameter, C_n^2. In particular for Southern India, Scintillation statistics (C_n^2) are estimated using Kolmogorov theory and Radiosonde data collected from India Meteorological Department observatory. The method proposed by Vasseur is applied for two years of Radiosonde Data collected at Bangalore (12.96N; 77.58E) which is the nearest radiosonde station to Hassan and the results are compared with observed scintillation values estimated form ITU-R method. C_n^2 values computed at Tirupati and SHAR centre, Sriharikota were also compared to understand the spatial and seasonal variation of C_n^2 values in southern India. Log-amplitude variation of scintillations was estimated for 15, 20 and 72 degree elevation angles and for 11.54, 19 and 30 GHz frequencies. It is observed from the Cumulative Distribution Function (CDF) that scintillation variance is higher for 19 GHz frequency and at low elevation angles. ITU-R estimated scintillation variance in dB is much lower than the observed values in probability range .01 to .001. This study clearly shows that even at high elevation angles in tropical countries the clear air scintillation variation cannot be ignored.

Most of the attenuation studies on earth-satellite paths have been carried out in the temperate regions of the world. The increasing use of satellites for telecommunications and broadcasting services have necessitated the need for earth-space attenuation studies in the tropics. In the tropical countries, rain is the main consideration, as the occurrence of other hydrometeors is very rare. The precipitation characteristics in the tropics differ appreciably from those of the temperate regions in that empirical relationships obtained in the latter may not be very suitable for system design in the former. An adequate knowledge of precipitation rate and rain attenuation is always required for a system designer to realistically determine link availability, establish the link margin and compensate for rain effects. Available rain measurements and the rain zones have not performed satisfactorily for rain attenuation calculations in the tropics. There are several factors contributing to this problem. For example, in the tropics, the rain drop size distribution is heavily biased towards larger drops for a given rate. This information can not be obtained from IMD data because of inadequate resolution in space and time. Yet another problem is due to differences in the melting layer height which is critical for earth-space paths.

Due to very few tropical rain attenuation measurements in existence, the available propagation models have not performed satisfactorily well in predicting the measured attenuation. Very little and sketchy information is available in India on these areas. In the past, attenuation measurements through radiometers forces one to look for the attenuation distributions derived from the satellite observations. With this in mind, preliminary studies are presented on rain attenuation using Indian satellite (INSAT-3B) signals at 11.54 GHz in *Chapter VII*. With the available rain data at the measurement site the Cumulative Distribution Function (CDF) for the signal strength and rain rate were estimated and the rain attenuation is estimated theoretically and compared with the experimental results.

It is observed that in general power spectral density of rain attenuation follows a slope of 20 dB^2 per decade and there are significant contributions from rain induced scintillations and clear air scintillation in observed signal power. Typical periodicity was observed in the rain fall structure on some days and correlated with the signal power attenuation. ITU-R predicted rain rate at 0.01 % probability closely matches with the observed rain fall at the experimental site but the attenuation provided by ITU-R is slightly lower than the observed attenuation.

The water vapor absorption at Ku- and Ka-bands is studied in *chapter-VIII.* Oxygen absorption is not considered, as the FSS (Fixed Satellite Services) do not employ the frequencies close to the oxygen absorption band (~60GHz) at present in India. Therefore, water vapor absorption alone was considered in this study. The attenuation of Satellite signals due to water vapor absorption was calculated using the formulas referred in Recommendations ITU-R P676-5 and compared with the water vapor attenuation calculated using Radiosonde data collected from the India Meteorological Department (IMD) observatory at Hyderabad. Radiosonde data was collected for a three year period starting from Jan 2002 to Oct 2004. Predictions of specific attenuation and path attenuation due to water vapor absorption are presented in this chapter. The yearly mean zenith water vapor absorption using the three years of radiosonde data has been computed at frequencies, f=15 GHz and at f=30 GHz and compared with the Recommendation: ITU-R P.676-5 predictions. It is found that our predictions are higher than the ITU-R P.676-5 predictions. The obvious reason for the differences is that the amount of water vapor content here significantly differs from ITU-R values. It is important to realize that water vapour absorption can be serious in earth-space paths above 20 GHz, but can cripple terrestrial links even at 15 GHz because the total path will be in the lower troposphere where in tropics the water vapour content can be higher than 30 g/m 3. At middle and high latitudes where temperatures rarely exceed 25°C, this problem is less critical. It may be of interest to note that even in torrential rains, the total water content does not exceed 10g/m^3.

POTENTIAL FOR FUTURE WORK IN THIS AREA

As HF Communications remain critical for many priority and marginal users in India and in other low latitude countries, the frequency management in this spectrum should be implemented in three stages. Firstly, the long-term prediction of cardinal ionospheric parameters should be given to users based on long-term solar activity predictions. Here, attempts should be made for the possible use of solar EUV data from satellites to predict ionospheric parameters rather than the surrogate sunspot number. Saturation effects that are usually associated with the usual solar indices such as sunspot number or 10.7 cm flux in the foF_2 values can be avoided with EUV indices if they are developed with all the available data. A very successful attempt was made in this direction by *Lakshmi et a.,l (1988)*, but more data need to be used for a good statistically significant indices to be evolved, secondly, short term forecasting should be improved (including solar flares, CME events and storms) so that the monthly medians provided in the long term forecast can be updated. Thirdly, a Real Time Channel Evaluation (RTCE) should be performed as the last stage especially for high speed digital data transfer.

In the area of micro-wave communications, both rain attenuation and water vapour attenuation are critical at low latitudes. While rain attenuation per km for a given rain rate is known with reasonable accuracy, rain rates are not available in many geographical regions with desired detail. Most of the data collected by the Meteorology depts. in tropics is only good for flood forecasting etc. and not for use in communication applications. Rain gauges with very short integration times like several seconds are required to define margins for high reliability digital links. Very comprehensive campaigns in countries like India are required and such planning requires considerable investments both in manpower and financial inputs.

The saga of high speed digital communications facilitated by advances in digital electronics and computers can be exploited only if the impairments caused by the propagation medium are understood, quantified and accounted for while designing high quality links. The work in this thesis is a modest attempt to high-light the issues through data collected from several places in India and it is hoped that this work will excite other workers to take up problems that are peculiar to low latitudes.

REFERENCES

Abdu, M.A., J.H.A., Sobral, P.Richards, M.de Gonzalez, Y,N.Haung, B.M.Reddy, K.Cheng, E.P.Szuszczewicz and I.S.Batista, *J. of geophys. Res.* Vol.101, No.A12, p 26, 729-26, 740, 1996.

Aggarwal, S., D.R. Lakshmi and B.M. Reddy, *Solar Terrestrial Prediction Proceedings,* edited by R.F. Donnelly, U.S. Dept. of Commerce, USA, 1, 134, 1979.

Aggarwal, S., and S. Shastry, *published by National Physical Laboratory,* New Delhi, 1991.

Agrawal R.A., *IEEE. Trans. Electromagn. Compact.,* EMC, 25 (a), 70, 1983.

Agrawal R.A., *IEEE. Trans. Electromagn. Compact.,* EMC- 25 (b), 119, 1983.

Agrawal R.A., *IEEE. Trans. Electromagn. Compac.,* EMC- 25 (c), 389, 1983.

Aiya, S.V.C., *J.Atmos. Terr. Phys.,* 5, 230, 1954

Aiya S.V.C., *J. Ind. Inst. Sci.* 5, 113, 1977.

Andersen, J.B and Pedersen, K.I., IEEE Trans, Antennas and propagation, 50(3), pp. 391-395, 2002.

Antonio Fischer De Toledo, Adel M.D. Turkmani and J. David Parsons, IEEE Personal Communications, p.40, April 1998.

Appleton, E.V., and Pigott, W.R., J. Atmos. *Terr. Physc.,* 9, 141, 1971.

Arns, J., *IEEE Trans. Antennas Propogat,* AP-25, 729, 1977.

Ashok Kumar, et al., *IEEE Trans. Ant. Propagation,* Vol.52, No.3, p.702-708, March 2004.

Ashok Kumar Baldotra and I.S. Hudiara, *IEEE Trans. Ant. Propagation,* Vol.52, No.6, p.1505-1508, June 2004.

Asoka Dissanayake; Jeremy Allnut, Fatim Haidara, *International Journal of Satellite Communications*, 19:p.335-345, 2001.

Asoka Dissanayake, Jeremy Allnut, Fatim Haidara, *Online Journal of Space Communication*, Issue No.2, p1-36, 2002.

Asoka Dissanayake, *Online Journal ofSpace Communication*, Issue No.2, p1-5, 2002.

Aurand, J.F., and Post R.E., *IEEE Trans. Veh.Technol*, Vol.37, p.149, 1988.

Barelay, P.A., Bennet J.A. and Boston R.C., *Report MEE-78m INTELSAT, Contract*, IS-830, 1978.

Basu, S., S.V.Basu, *COSPAR symposium series* 8: p187-199. Perganon Press, 1980.

Bernard De Backer etal, IEEE Trans. Veh. Techn., 52, 5, pp. 1297-1301, Sept 2003.

Blomquist, A., and Ladell L, *FOA Report*, 1974.

Boris Christain Gremont and Miodrag Filip, *IEEE Trans. Ant. Propagation*, Vol.52, No.5, p.1245– May 2004

Brussard, G., Dijk, J., and Wijdemans L., *Eindhoven University of Technology and Surabaya Institute of Technology, Final Report, Eindhoven*, 1993.

Capsoni, C., Fedi, F., and Watson, P.A., *Radio Sci.*, Vol.22, p.387-393, 1987.

CCIR, *World Distribution and Characteristics of Atmospheric Radio Noise (International Telecommunication Union), Geneva, CCIR Report* No.–322 1963

CCIR (2-30 MHz), Report 252, *International Telecommunication Union, Geneva*, 1970.

CCIR *Documents a. 6/88 (Federal Republic of Germany)*; b.10/56 (India); c.10/323 (India), ITU, Geneva, 1974-78.

CCIR *Report 252-2 (supplement), ITU*, Geneva, 1978.

CCIR, *Documents* a. 6/77 (India); b. 6/76 (India), ITU, 1978-82.

CCIR, *Rep. No. 322-2*, 1985 .

CCIR *Recommendations and Reports of the CCIR*, Vol.6, 239, ITU, Geneva, 1986.

CCIR, *9th Plenary Assembly of CCIR (International Telecommunication Union, Geneva)*, Rep.No. 303-1, 183, 1966.

Chandrasekhar, V., *IEEE Trans. Geoscience and Remotesensing,* Vol.41, No.10, p.2166 - 2176, Oct.2003

Cost 231., *Final Report, European Commission,* 1999.

Crane, R.K., *IEEE Trans Comm.* Vol.28, No.9, p.1717-1733, 1980.

Crane, R.K., *Proc. IEEE,* Vol.69, No.2, 1981.

Crane, R.K., *Radio Sci.,* Vol.20, No.4, p.843-863, 1985.

Crane, R.K., *John Wiley & Sons,* New York, 1996.

Costa, R.A., S.Johnson, W.Feliciano, R.Pollard, L.A.Gonzalez *"NASA'S New Radio Propagation Experiment"*, Website, http://spacecom.grc.nasa.gov/projects/mmw/8thka_paper.pdf

Dabas, R.S., Lakshmi, D.R., and Reddy, B.M, *Radio Science* 24, p563-573, 1989.

Davies, P.G., and Mackenzie E.C., *IEE Proc.* Vol.128, Pt.H, No.1, p.53-65, 1981.

Dissanayake, A., Allnutt J and McCarthy D.K., *Electronic Letts.,* Vol.26, No.3, p.168-170, 1990.

Egli, J.J., *Proc. IRE,* Vol.45, pp.1383-1391, 1957.

Fedi, F., *Alta Frequenza,* Vol.66, p.167-184, 1979.

Ganalingam,S., Ahmedabad (Physical Research Laboratory, Ahmedabad), Vol.1, 47. 1969.

Gareia Lopez, J.A., Hernando J.M. and Selga J.M., *IEEE Trans. Ant. Prop.,* Vol.36, No.3, p.444, 1988.

George, P.L., J.Atmos. *Terr. Phys.,* 32, 1893-1906, 1971.

Ghosh & Mitra S.N., and Ghosh, B.B., *J. Instn. Telcomm. Engrs.,* New Delhi, 52, 1958,

Ghosh, B.B., and Saksena, R.C., *"Indian J. Radio Space Phys"*, 7 (1978), 205.

Huang Y.N., Radio Sci., 12, 239, 1977.

Ghosh, B.B., and Mitra, S.N., *J. Instn. Telecomm. Engrs.,* New Delhi, 5,194, 1959.

Ghosh, B.B., Saksena, R.C., and Malhotra, *V.J., Indian J. Radio and Space Phys.,* 3, 37, 1974.

Ghosh, B.B., Chakravarty, D.K., and Malhotra , *V.J., Indian J. Radio and Space Phys.* 4, 104, 1975.

Ghosh, B.B., Saksena, R.C., and Malhotra V.J., *IJRSP,* 3 , 37, 1974.

Ghosh, B.B., Saksena, R.C., and Malhotra, V.J., 1978, *IJRSP*, 7, 106

Grosskopf, R., *IEEE Trans.Ant. Propagation*, Vol. AP-35, No.7 pp.852-859, 1987.

Gunn, R., and Kinzer G.D., *J. Meteorology*, Vol.6, p.243-248, 1949.

Hall, M.P.M., *Peter Pereperegrinus Ltd., (inst. Electrical Engrs.)*, 1979.

HAPS Reference - Website http://www.bakom.ch/imperia/md/content/english/funk/forschungundentwicklung/studien/HAPS.pdf

Hata, M., *IEEE Trans. Veh. Technol.*, Vol.29, p.317-325, 1980.

Herman, J.R., *J. Res. Natn. Bur. Stand.*, 65D, 565, 1961.

H.L. Bertoni, Radio Propagation for modern wireless Systems. Upper Saddle River, NJ: Prentice-Hall, 2000.

Hogg, D.C., and Chu T., *Proc. IEEE*, Vol.63, p.1308-1331, 1975.

Howard Sandler, Nortel Networks Corporation, 100 Constellation Cr, Nepean, Ontario, K2G 6J8,Canada, IEEE 802.16cc-99/15, Broadband Wireless Access Working Group, *Propagation Model for Coexistence Modelin*, 1999.

Ibukin, O.J., Atmos, *J., Terr. Phy.*, 27, 1059, 1965.

Ippolito, L.J., *Proc. IEEE*, Vol.69, 1981.

Ippolito, J.L., *Proc. IEEE*, Vol.69, p.697-727, 1981.

Ippolito, J.L., *Van Nostrand Reinhold Company*, New York, 1986.

ITU-R / CCIR., *Recommendations and Reports of the CCIR 1990*, Vol. VI, Report No.266-7, ITU-R Geneva, 1990.

ITU-R *recommendation*, P.525-2, 1994.

ITU-R *recommendation*, P.529-1, 1994.

ITU-R *recommendation*, P.837-1, 1994.

ITU-R *recommendation*, P.529-2, 1995.

ITU-R *recommendation*, P.530-8, 1997

ITU-R *recommendations*, P.618-5, 1997.

ITU-R *recommendations*, P.839, 1997.

I ITU-R *recommendations* P.835-3 [14], (Geneva), 1999

ITU-R *recommendations*, P.529-3, 1999.

ITU-R *recommendations* P.453-8, *(Geneva), 2001.*

ITU-R *recommendations*: P.676-5, *(Geneva), 2001.*

ITU-R *recommendations,* P.839-3, 2001.

Joglekar, P.J., *J. Inst. Telecomm. Engrs.*, New Delhi 17 (1971), 92.

Joglekar, P.J., and T.S. Sathiamurthy, *Radio Sci.*, 10, 601, 1975.

Joss, J., Thomas, J.C., and A.Waldvogel, *Proc. Intl. Conf. Clond Phys.*, (Toronto), 1968.

Kim, S.O., et al., IEEE Trans. Veh. Techn., 48(3), pp. 931-945, May 1999.

Kolomogorov, A.N., *C.R.Academy of sciences,* URSS, 30, 1941.

Kotaki, M., Kuriki I., Katoh C. and Sugiuchi H, *J. Rdio Res. Lab.*, 28 49. 1981,

Kotaki, M., and Katoh C., *J. Radio Res. Lab.*, 30 ,35, 1983.

Kotaki, M., and Katoh C., *J. Atmos. Terr. Phys.*, 45, 834, 1983.

Kotaki, M., J. Atmos, *J. Terr. Phys.*, 46, 867, 1984.

Kurner,T., et al., IEEE Journal on selected areas in communications, 111(7), pp. 1002-1012, 1993.

Kurner, T., Proc. VTC 99, Houston, 1999.

Lakshmi, D.R., Aggrawal, S., Paricha, P.K., & Reddy, B.M., *J. of Radio & Space Physics,* 8, 101-105, 1979.

Lakshmi, D.R., Reddy, B.M., Chakraborthy, R. and Mangal Sain ,*Proc. Indian National Science Academy,* 48 A, Supplement No.3, 308, 1982.

Lakshmi, D.R., Mangal Sain, Iqbal Ahmed, M. M. Gupta, and B. Veenadhari, *IEE* 392, 91, 1994.

Laws, J.O., and Parsons D.A., *Trans. Am. Geophys. Union.*, Vol.24, p.452-460, 1943.

Laws, J.C., and D.A.Parsons, *Trans. American Geophys. Union,* Vol.24, 1943.

Lee, W.C.Y., *McGraw Hill,* New York, 1998.

Lekkla, R., Lim, S.L., Wachja, J., and McCormick K.S., *Proc. URSI Open Symp. Climpara* 194, 5.12.1-5.12.4, 1994.

Mashall, J.S., and W.M Palmer, *The Distribution of Rain Drops with Size, J.meteor,* Vol.5, 1948.

Manning, R.M., *Int. J. Sat. Comm.*, Vol.2, p.187-197, 1984.

Matricciani, E., *IEEE Trans. Broad.*, Vol.44, No.2, p.250-258, 1998.

Matricciani, E., Mario Mauri and Riva C., *Radio Science,* Vol.32, No.5, p.1861-1866, 1997 (a).

Matricciani, E., *Radio Science,* Vol.32, No.3, p.935-941, 1997(b).

Matricciani, E., *Electron Lett.*, Vol.30, No.6, p.522-524, 1994.

Matricciani, E , IEEE Trans. Ant propagation, Vol.52, No.8, p.2083 –August 2004

Mc Carthy, D.K., Allnutt J.E., Salazar W.E., Wanmi F., Tchinda M., Adinayi T.D.G and Zaks.Co., *Electronic Letts.*, vol.28, no.3, p.290-292, 1992(a).

Mc Carthy, D.K., Allnutt J.E., Salazar W.E., Omeata E.C., Owolabi B.R., Oladiran T., Ajayi G.O., Raji T.1 and Zakz C *Electronic Letts.*, Vol.28, No.3, p.318-319, 1992(c).

Mondal , N. C, K I Timothy and S K Sarkar, Intl J Remote Sensing, 18,3669-3675, 1997

Mondal, N.C, S K Sarkar, International Journal of Infrared and Millimeter waves, (USA), 22, 495-503, 2001,

Mondal, N.C, S K Sarkar, Theoretical and Applied Climatology, (USA), 76/1-2, 89-104, 2003

Moupfouma, F., *Proc. URSI General Assembly,* Toronto, Canada, FP.28, p.380, 1999.

Moupfouma, F., *Ann. Telecommun,* Vol.42, No.9-10, p.539-550, 1987 (a).

Moupfouma, F., *IEE Proc.,* Vol.134, Pt.H, No.6, p.527-537, 1987(b).

Moupfouma, F., *IEEE Trans. Ant. Prop.*, Vol.32, No.12, p.1368-1372, 1984.

Muller, E.A., and Jones D.M.A., *Proc. 8th Weather Radar Conf. Boston,* USA, 1960.

Nackoney, O.G., and Dvidson D., *Radio Sci.*, Vol.17, No.6, 1435-1442, 1982.

Narayana Rao., M.J.kesava Murthy, S.K.Sarkar, H.N.Dutta amd B.M.Reddy *Ind.J.Radio & Space Physics,* 14,83, 1985.

Oguchi, T., *Proc. IEE* Vol.71, No.9, p.1029-1078, 1983.

Okumara, Y., Ohmori E, Kawano T and Fukuda K., *Rev. Elec. Commun Lab.*, Vol.16, p-825-873, 1968.

Ong, J.T., etal., Electronics Letters, Vol.36, No.7, p.661-663, March, 2000

Oort, A.H., and Rasmusson E.M., *NOAA Professional paper 5* (US Dept. of Commerce), 1971.

Ottersten, H., *Radio Sci.*, 4, 1251-1255, 1969.

Peng Xu and Leung Tsang, IEEE Antennas and Wireless propagation letters, 3, p.336-339, 2004.

Poloniȯ, R., and Riva C, *IEEE Trans. Ant. Prop.*, Vol.46, No.5, p.631-635, 1998.

Pontes, M.S., Mello L.A.R.S., Souza R.S.L and Migliora *C.G.S., Int. J. Sat. Comm.*, Vol.8, No.3, p.239-249, 1990.

Prasad, M.V.S.N., and Ahmed I., *IEEE Trans. Broad.*, Vol.43, No.4, p.459-486, 1997.

Prasad, M.V.S.N., and Sarkar S.K., Sain M and Reddy B.M., *Indian J. Rad, & Space Phy.*, Vol.23, p.135-141, 1994.

Prasad, M.V.S.N., and Rajendra Singh, IEEE Trans. Veh. Techn., 52, 3, p.671-681, May 2003.

Prasad, M.V.S.N., and Rajendra Singh, IEEE Trans. Veh. Techn., 49,1, p.239-245, January 2000.

Press, W.H., B.P.Flannery, S.A.Teukolsky, and W.T.Vetterling, Cambridge, U.K., *Cambridge Univ., Press,* 1988.

Rama Rao, P.V.S., B.V.P.S. Rao and A. Maitra., Ind. *J. Radio & Space Phys.*, 9, 63, 1980.

Ramachandran, V and Vickal Kumar, S.pac.J. Nat. Sci, 21, p.59-62, 2003

Raina, M.K., and Uppal G.S., *IEEE Trans. Ant. Prop.*, vol.29, no.6, p.857-864, 1981.

Rastogi, R.G., *J A T P*, 42, 593, 1980.

Ray, P.S., *Applied Optics,* Vol.11, No.8, p.1836-1844, 1972.

Rama Rao, P.V.S., *Ionosonde data of* February 1995 and November 1997-April 1998.

Rama Rao P.V.S., D.Nru and M. Srinivasa Rao, *Proceedings of COSPAR/URSI symposium,* 175, 1981.

Rao, M.K., Muzamdar, S.C and Mitra, S.N. *J.Atoms. Terr. Phys.* 24,245, 1962

Rao, M.K. *J.Inst. Electronics and Telecom. Engrs.* 15, 729, 1969.

Rateliffe, J.A., *Cambridge University Press, Cambridge,* 1959.

Reddy, B.M., Aggarwal, S., Lakshmi, D.R., Shastry, S. and Mitra, A.P., *Solar-Terrestrial Prediction proceedings, edited by R.F.Donnelly* (US Deptt. Of Commerce, USA), 1, 119, 1979.

Reddy, B.M., *Hand book on Radio Propagation for Tropical and Sub-tropical Countries, published by URSI Committee on developing countries,* 1985.

Rodda, M.J., and Williamson A.G., *Electronic Letts.,* Vol.32., p.397-399, 1996.

Rush, C.M., D. Miller and J. Gibbs, *Radio Science,* 9, 749. 1974.

Sarkar, S.K., H.N.Dutta, P.K.Pasricha and B.M. Reddy, *published by National Physical Laboratory,* New Delhi, 1982.

Samuel, J.C., and Bradley, P.AJ. *Atmos. Terr. Phys.,* 37, 131-141, 1975.

Sailors, D.B., and Brown R.P., *Radio Sci.,* 18 625, 1983.

Satyam, M., *J. Sci. Indust. Res.,* 21D 221, 1962.

Sakesena, R.C., *Indian J. Radio & Space Phys.,* 3, 318, 1975.

Sarkar, S.K., P.K. Paricha, H.N. Dutta, B.M. Reddy and S.M. Kulshrastha, *published by National Physical Laboratory,* New Delhi, 1985.

Sarkar, S.K., Ahmed I., Prasad M.V.S.N., Dutta H. and Reddy B.M., *reference data manual, National Physical Laboratory,* New Delhi, 1992.

Sarkar S K, N C Mondal International Journal of Satellite communication, (UK) 16, 127-130, 1998

Sarkar, S. K, Rajesh Kumar and J Das, Indian Journal of Physics, 77B, 271-275, 2003

Sarkar, S. K, R Kumar, I Ahmad, M M Gupta, M V S N Prasad, J Das and A K De, Indian Journal of Space Physics, 33, 176-179, 2004

Sarkar, S. K, Iqbal Ahmad, M M Gupta, M V S N Prasad, J Das, A K De and Rajesh Kumar, Indian Journal of Physics, 79, 117-121, 2005

Sarkar, S.K, Iqbal Ahmad, J Das and A K De, International Journal of Infrared and Millimeter waves, (USA), 26, 329-340, 2005

Sarma, A.D., G. Sasibhushana Rao and V. Venkata Rao *Ionospheric Reference.*

Sen, H.K., and Wyller, A.A. J. Geo Phys. Res., 65, 3931, 1960.

Sangupta.N., Turney O and Brussard G., *IEE Proc.* 134, Pt.F, p.583-596, 1987.

Station Placement for INWAAS-A preliminary study" J. Indian Geophys. Union, Vol-4, No.1, 41-49, 2000.

Shinji Jinguu and Hiroaki Nakabayashi, IEEE Trans. Veh. TEchn, 52, 1, p.108, January 2003.

Steele, R., *IEEE Press & Pentech Press,* 1992.

Sehgal, O.P., Agrawal, H.O., Mangal Sain and Rao, M.K. *J. Inst. Electronics and Telecom. Engrs.,* 28 (2) 68, 1982.

Takeshi Hatsuda et.al., IEEE Tran. Ant & Propagation, Vol.52, No.6, p.1481-1491,

June 2004.

Tatarskii, V.I., *McGraw Hill Book Co., Inc.* New York, 1961.

Timothy, K.I. etal., Electronics Letters, Vol.31, No.17, p.1505-1506, August 1995

Timothy, K.I and S.K. Sakar, Electronics Letters, Vol.33, No.10 p.895-897, May 1997.

Timothy,K.I, N.C.Mondal and S.K.Sarkar, International Journal of Satellite

Communication,(UK), 16, 53-57, 1998

Timothy, K.I. etal., Electronics Letters, Vol.36, No.10, p.894-895, May 2000.

Timothy, K.I et.al., IEEE Communication Letters, Vol.5, No.2, p.49-51, Feb.2001.

Ulbrich, C.W., *J. Clim. And Appl. Meteorology,* p.1764-1775, 1983.

Van de Kamp, M.M.J.L., Rica C., Tervonen J.K and Salone E.T., *IEEE Trans. Ant. Prop.,* Vol.47, No.1, p.77-85, 1999.

Vande Kamp, M.M.J.L and Castanet, L., Fade Dynamics Review, COST 280, PM 3018, Ist International workshop, p.1-20, July 2002

Vasseur, H., *IEEE Trans. Ant. Prop.,* Vol.47, No.2, p.293-301, 1999.

Vijay, K.G .,and Joseph E.W., *Prentice Hall,* 1996.

Vogel, E.J., Torrence, G.W and Allnutt, J.E., *Proc. IEEE,* Vol.81, p.885-896, 1993.

Wakai, N., *J. Radio Res. Labs,* Japan, 18 (37), 213, 1961.

Wakai, N., J. Radio Res. Labs. Japan, 18 (98), 245, 1971.

Walfisch, J., and Bertoni, H.L., *IEEE Trans. Ant. Prop.*, Vol.36, No.12, pp1788-1796, 1988.

Waters, J.W., Vol.12B, M.L.Meeks, Ed., *New York Academic Press,* 1976.

Watson, P.A., Leitao M.J., Sathiaseelan V., Gunes G., Baptista J.P.V.P., Potter B.A.,

Wei Zhang, IEEE Transactions on Antennas and Propagation, Vol. 45, No. 11, 1669, 1997.

Winn, W., Modelling, Review of Radio Science Edited by Ross Store. 1996-1999, Winn ,W., W.P., Schwede G.W. and Moore C.B., J. *Geo. Phy. Res.,* 79, 1761, 1974.

Yamada, M., and Miura Y., *Microwave and Millimeter Wave Technology Proceedings, ICMMT'98,* p.1016-1019, 1998.